Javad Lavaei
Amir G. Aghdam

Decentralized Control of Interconnected Systems

Javad Lavaei
Amir G. Aghdam

Decentralized Control of Interconnected Systems

VDM Verlag Dr. Müller

Imprint

Bibliographic information by the German National Library: The German National Library lists this publication at the German National Bibliography; detailed bibliographic information is available on the Internet at http://dnb.d-nb.de.

Cover image: www.purestockx.com

Publisher:
VDM Verlag Dr. Müller Aktiengesellschaft & Co. KG , Dudweiler Landstr. 125 a, 66123 Saarbrücken, Germany,
Phone +49 681 9100-698, Fax +49 681 9100-988,
Email: info@vdm-verlag.de

Zugl.: Montreal, Concordia University, Diss., 2006

Produced in USA and UK by:
Lightning Source Inc., La Vergne, Tennessee, USA
Lightning Source UK Ltd., Milton Keynes, UK
BookSurge LLC, 5341 Dorchester Road, Suite 16, North Charleston, SC 29418, USA

ISBN: 978-3-639-03417-2

Contents

Chapter 1

Introduction

Control of interconnected systems has been of great interest in the literature in the past three decades, due to its numerous applications in important real-world problems. Such applications include for example power systems, communication networks, flexible space structures, to name only a few. Due to the distributed nature of the problems of this type, the conventional control techniques are often not capable of handling them efficiently. More specifically, it is desired in the distributed interconnected systems to impose some constraints on the structure of the controller to be designed. These constraints specify the outputs of which subsystems can contribute to the construction of the input of any subsystem. To formulate the control problem, these constraints are usually represented by a matrix, which is often referred to as the information flow matrix.

A special case of structurally constrained controllers is when the controller of each subsystem operates independently of the other subsystems; i.e. when there is no direct interaction between the control effort of each subsystem and the output signals of the other subsystems. This case is of a particular interest in the control literature, and is usually referred to as *decentralized* control. Each control component in a decentralized control system observes only the output of its corresponding subsystem to construct the input of that subsystem. The notion of a decentralized fixed mode (DFM) was introduced in the literature to characterize the modes of an interconnected system which are fixed with respect to any decentralized linear time-invariant (LTI) controller. Since a DFM may not be fixed with respect to a nonlinear and time-varying controller, the notion of a quotient fixed mode (QFM) was introduced later on to identify those modes that are fixed with respect to any type of decentralized control law (i.e., nonlinear and time-varying). Various properties of decentralized controllers are investigated thoroughly in the literature.

More recently, the case when the local controllers of an interconnected system can communicate with each other has been studied intensively in the literature. This problem is referred to as *decentralized overlapping control*, and is motivated by the following practical issues:

1. The subsystems of many interconnected systems (referred to as overlapping subsystems) share some states. In this case, it is often desired that the structure of the controller matches the overlapping structure of the system.

2. Sometimes in centralized control systems, there are limitations on the availability of the states. In this case, only a certain subset of outputs are available for constructing each control signal. However, the control structure is not necessarily localized.

This book aims to investigate different aspects of *decentralized* and *decentralized overlapping* control designs, such as stabilization, optimality, and robustness. To this end, eight chapters are included following this chapter to investigate a number of incorporated problems step by step. The relevant subjects are spelled out below.

First, in Chapter 2 the problem of designing a near-optimal decentralized control law for acyclic interconnected systems is studied and compared to the existing methods. Robustness of the proposed control law is also investigated to verify its practical applicability, and its robust performance is evaluated accordingly. The proposed method is applied to a formation of three vehicles, which manifestly demonstrates its efficacy.

Next, the technique exploited to design a near-optimal decentralized controller for acyclic systems is further developed in Chapter 3 for the case of general interconnected systems. It is shown that any given centralized controller can be equivalently transformed into a decentralized one at the cost of more computational effort. A simple procedure is presented to design a decentralized controller with the aim of achieving some desirable objectives, based on the available centralized techniques. The key features of the proposed control law are studied accordingly.

It is known that discrete-time decentralized controllers can potentially outperform their continuous-time counterparts in a broad class of interconnected systems. The problem of designing a decentralized generalized sampled-data hold function for interconnected systems is considered in Chapter 4. Some recent results in this area are utilized to solve the problem of global optimization of a rational function. The proposed design technique has proved to be quite efficient and superior to the existing work.

The method developed in the preceding chapter for designing a periodic controller is suitable for medium-sized systems. Thus, a linear matrix inequality (LMI) based technique is introduced in Chapter 5 which can be applied to large-scale systems and has significant advantages. More precisely, for a given set of LTI systems, a periodic controller along with a compensator is proposed to stabilize all the systems simultaneously, while it acts as an optimal controller for each individual system.

By virtue of the restrictions in decentralized control, a system might have some DFMs which are not movable via LTI decentralized controllers (as pointed out earlier). These modes deteriorate the performance of the control system, and may lead to instability. The question arises: How can these modes be eliminated? It is shown in Chapter 6 that the distinct and nonzero DFMs of a system can always be eliminated by means of sampling. This means that sampled-data decentralized controllers can be deployed to control the systems that may not be stabilizable by means of the conventional continuous-time LTI decentralized controllers.

As mentioned earlier, the fixed modes of any system play a crucial role in control design. Thus, characterization of fixed modes is intensively investigated in the literature. Since the existing analytical methods for finding fixed modes are generally ill-conditioned and computationally inefficient, a novel graph-theoretic approach is introduced in Chapter 7 to obtain DFMs in a more efficient manner. This technique immensely diminishes the computational cost, and is attractive from several standpoints, as discussed in this chapter. Moreover, the proposed approach is extended to identify the QFMs of the system, which are the fixed modes of the system with respect to any decentralized control law (i.e., nonlinear and time-varying).

The methods proposed in the preceding chapters for high-performance continuous and discrete feedback designs are mainly for the case when the control structure is strictly decentralized (i.e., when the control configuration is localized). It is shown in Chapter 8 how these results can be extended to the case of decentralized overlapping control design. The notions of a decentralized overlapping fixed mode (DOFM) and a quotient overlapping fixed mode (QOFM) are introduced. The significance of these notions as well as their relevance to DFMs and QFMs are discussed thoroughly.

In practice, the controllers obtained in the preceding chapters (either strictly decentralized, or decentralized overlapping) are to be applied to the systems subject to parameter uncertainty and perturbation. For a more pragmatic view, it is assumed in Chapter 9 that the closed-loop system is polynomially uncertain. Some important results are obtained for the robust stability verification of the system in this case. The result obtained presents the first necessary and sufficient condition in the literature for the corresponding robust stability problem. This condition is in the form of sum-of-squares (SOS).

It is important to note that parts of this book have been previously published in some journal publications and, therefore, their usage is subject to the copyrights of the corresponding publications. The list of publications relevant to this book are given below:

- Javad Lavaei, Ahmadreza Momeni and Amir Aghdam, "Spacecraft Formation Control in Deep Space with Reduced Communication Requirement," *IEEE Transactions on Control Systems Technology*, vol. 16, no. 2, pp. 268-278, Feb. 2008 (Chapter 2).

- Javad Lavaei and Amir Aghdam, "High-Performance Decentralized Control Design for General Interconnected Systems with Applications in Cooperative Control," *International Journal of Control*, vol. 80, no. 6, pp. 935-951, Jun. 2007 (Chapter 3).

- Javad Lavaei and Amir Aghdam, "Optimal Periodic Feedback Design for Continuous-Time LTI Systems with Constrained Control Structure," *International Journal of Control*, vol. 80, no. 2, pp. 220-230, Feb. 2007 (Chapter 4).

- Javad Lavaei and Amir Aghdam, "Simultaneous LQ Control of a Set of LTI Systems using Constrained Generalized Sampled-Data Hold Functions," *Automatica*, vol. 43, no. 2, pp. 274-280, Feb. 2007 (Chapter 5).

- Javad Lavaei and Amir G. Aghdam, "Stabilization of decentralized control systems by means of periodic feedback," *Automatica*, vol. 44, no. 4, pp. 1120-1126, Apr. 2008 (Chapter 6).

- Javad Lavaei and Amir Aghdam, "A Graph Theoretic Method to Find Decentralized Fixed Modes of LTI Systems," *Automatica*, vol. 43, no. 12, pp. 2129-2133, Dec. 2007 (Chapter 7).

- Javad Lavaei and Amir Aghdam, "Control of Continuous-Time LTI Systems by Means of Structurally Constrained Controllers," *Automatica*, vol. 44, no. 1, Jan. 2008 (Chapter 8).

- Javad Lavaei and Amir G. Aghdam, "Robust Stability of LTI Systems over Semi-Algebraic Sets using Sum-of-Squares Matrix Polynomials," *IEEE Transactions on Automatic Control*, vol. 53, no. 1, pp. 417-423, Feb. 2008 (Chapter 9).

Chapter 2

A Near-Optimal Decentralized Control Law for Interconnected Systems

2.1 Abstract

In this chapter, an incrementally linear decentralized control law is proposed for the formation of co-operative vehicles with leader-follower topology. It is assumed that each vehicle knows the modeling parameters of other vehicles with uncertainty as well as the expected values of their initial states. A decentralized control law is proposed, which aims to perform as close as possible to a centralized LQR controller. It is shown that the decentralized controller behaves the same as its centralized counterpart, provided *a priori* information of each vehicle about others is accurate. Since this condition does not hold in practice, a method is presented to evaluate the deviation of the performance of the decentralized system from that of its centralized counterpart. Furthermore, the necessary and sufficient conditions for the stability of the overall closed-loop system in presence of parameter perturbations are given through a series of simple tests. It is shown that stability of the overall system is independent of the magnitude of the expected value of the initial states. Moreover, it is shown that the decentralized control system is likely to be more robust than the centralized one. Optimal decentralized cheap control problem is then investigated for the leader-follower formation structure, and a closed form solution is given for the case when the system parameters meet a certain condition. Simulation results demonstrate the effectiveness of the proposed controller in terms of feasibility and performance.

2.2 Introduction

In the past several years, a certain class of interconnected systems, namely acyclic systems, has found applications in different practical problems such as formation flight, underwater vehicles, automated highway, robotics, satellite constellation, etc., which have leader-follower structures or structures with virtual leaders [1-13]. The main feature of this class of systems is that their structural graphs are acyclic, i.e. they do not have any directed cycles.

In a leader-follower formation structure, each vehicle is provided with some information (e.g., acceleration or velocity) of certain set of vehicles. It is shown in the literature that the control problem of such formation can be formulated as the *decentralized* control problem of an acyclic interconnected system, where each local controller uses only the information of its corresponding subsystem (e.g., see [2]). The objective of this chapter is to design a decentralized controller which stabilizes any system with an acyclic structure, and performs sufficiently close to the optimal centralized controller. In other words, it is desired to reduce the degradation of the performance due to the information flow constraints in decentralized control structures.

During the past three decades, much effort has been made to formulate the optimal decentralized control problem, or solve it numerically. The main objective is to find a decentralized feedback law for an interconnected system in order to attain a sufficiently small performance index. These works can mainly be categorized as follows:

1. The first approach is to eliminate all of the interconnections between the subsystems in order to obtain a set of decoupled subsystems, and then design a local optimal controller for each of the resultant isolated subsystems [13], [14], [15]. Since the effect of interconnections has been neglected in this design procedure, the resultant closed-loop system with these local controllers may be unstable. Even when the interconnected system under the above controllers is stable, the performance index may be poor. As a result, this decentralized control design technique is ineffective in presence of strong coupling between the subsystems.

2. Another approach is to obtain a decentralized static output feedback law by using iterative numerical algorithms in order to minimize the expected value of the quadratic performance index with respect to an initial state with a given probability distribution [16], [17], [18], [19]. This type of design techniques are, in fact, the extended versions of the algorithms for designing

optimal centralized static output feedback gain, such as Levine-Athans and Anderson-Moore methods. Although these techniques result in a better performance compared to the preceding method, they have several shortcomings. First of all, they only present necessary conditions, which are mainly in the form of complicated coupled nonlinear matrix equations. Secondly, these iterative algorithms require an initial stabilizing static gain, which should satisfy some requirements. Finally, using a dynamic feedback law instead of a static, the overall performance of the system can be improved significantly.

3. In the third method, the optimal decentralized control problem is formulated by first imposing some assumptions to parameterize all decentralized stabilizing controllers, and then choosing the control parameters such that a desired performance is achieved [20], [21]. However, the resultant equations are either some sophisticated differential matrix equations or some nonconvex relations, which makes them very difficult to solve, in general.

4. This approach deals with a system with a hierarchical structure. For this class of systems, a rather centralized controller is designed in [22]. The decentralized version of this work is discussed in [23] for a discrete-time system. However, this method can analogously be applied to a continuous-time system, which can be interpreted as follows. In the hierarchical structure, consider the subsystem with the highest-level. Design a centralized local controller for that subsystem assuming that the remaining subsystems are in the open loop, which is desired to account for the performance index. Next, identify the second subsystem with the highest-level, and similarly, design a centralized local controller for it assuming that the first controller designed is a part of the system. Continuing this procedure, one can design all local controllers one at a time. The advantage of this method compared to the Method 2, explained above, is that it reduces off-line computation. However, this approach is inferior to Method 2, because the static gains are computed one at a time in this approach, while in Method 2, explained above, all of the static gains are determined simultaneously. As a result, this approach is proper once the order of the system is so high that the computational complexity is a crucial factor.

(note that the first three approaches are for general interconnected systems, while the last one is only for hierarchical systems). There are some other design techniques which are, in fact, combinations of methods 2 and 3 discussed above. All of these approaches are generally incapable of designing a decentralized controller with a satisfactory performance for most systems, including the class of interconnected systems with acyclic structural graphs.

This chapter presents a novel design strategy to obtain a high-performance decentralized control law for interconnected systems with leader-follower structure. It will later be shown that the proposed control law outperforms the first, the second and the fourth methods discussed above, and also has a simpler formulation compared to the third method. It is assumed that the state of each subsystem is available in its local output (this is a realistic assumption in many vehicle formation problems, e.g. see [24]), and that a quadratic cost function is defined to evaluate the control performance. The local controller of each subsystem is constructed based on *a priori* information about the model and initial states of all other subsystems. It is shown that if *a priori* knowledge of each subsystem is accurate, the performance of the decentralized control system is equal to the minimum achievable performance (which corresponds to the LQR centralized state feedback). In addition, a procedure is proposed to evaluate the closeness of the performance index in the decentralized case to the best achievable performance index (corresponding to the centralized LQR controller) in terms of the amount of inaccuracy in *a priori* knowledge of any subsystem. This enables the designer to statistically assess the performance of the proposed controller. Moreover, a set of easy-to-check necessary and sufficient conditions for the stability of the decentralized closed-loop system is given for both cases of exact and perturbed models for the system. It is to be noted that providing some information about the model of other subsystems for each individual local controller is performed off-line, in the beginning of control operation, and does not require any communication link between different subsystems. In other words, the proposed control structure is truly decentralized. Optimal cheap control problem is also studied for the leader-follower formation flying. This may require new actuators to be implemented on the vehicles in order to meet a condition on the input structure, which is necessary for the development of the results. While cheap control strategy may not be for many formation flying applications, e.g., constellation of satellites, where it is more desired to apply a minimum fuel control strategy, it can be very useful in certain formation applications involving UAVs with fast tracking missions. Throughout this chapter, each vehicle in formation will be referred to as a subsystem and the whole formation consisting of the leader(s) and all followers will be referred to as the system.

2.3 Decentralized control law

Consider a stabilizable interconnected system $\mathscr{S}(\mathscr{S}_1, \mathscr{S}_2, ..., \mathscr{S}_\nu)$ with the following state-space equation:

$$\dot{x} = Ax + Bu \tag{2.1}$$

where

$$A := \begin{bmatrix} A_{11} & 0 & \dots & 0 \\ A_{21} & A_{22} & \dots & 0 \\ \vdots & \vdots & \ddots & \vdots \\ A_{\nu 1} & A_{\nu 2} & \dots & A_{\nu\nu} \end{bmatrix}, \quad B := \begin{bmatrix} B_1 & \dots & 0 \\ \vdots & \ddots & \vdots \\ 0 & \dots & B_\nu \end{bmatrix} \quad u := \begin{bmatrix} u_1 \\ \vdots \\ u_\nu \end{bmatrix}, \quad x := \begin{bmatrix} x_1 \\ \vdots \\ x_\nu \end{bmatrix} \tag{2.2a}$$

and where $x_i \in \Re^{n_i}$, $u_i \in \Re^{m_i}$, $i \in \bar{\nu} := \{1, 2, ..., \nu\}$, are the state and the input of the ith subsystem $\mathscr{S}_i$, respectively. It is to be noted that the matrices A and B are block lower triangular and block diagonal, respectively. Assume that the state of each subsystem is available in the corresponding local output. This is a feasible assumption in most formation flying applications. For instance, the autonomous formation flying sensor (AFF) or the laser metrology [24] can be used to accurately measure the relative position and velocity in different formation flying missions.

Remark 1 *Consider an interconnected system whose structural graph is acyclic. It is known that the subsystems of this interconnected system can be renumbered in such a way that its corresponding matrix A is lower block diagonal [27]. In other words, any system with an acyclic structural graph can be converted to a system of the form $\mathscr{S}$ given by (2.1) and (2.2), by simply reordering its inputs and outputs properly, if necessary.*

Consider now the following quadratic performance index:

$$J = \int_0^\infty \left(x^T Q x + u^T R u\right) dt \tag{2.3}$$

where $R \in \Re^{m \times m}$ and $Q \in \Re^{n \times n}$ ($n := \sum_{i=1}^{\nu} n_i$, $m := \sum_{i=1}^{\nu} m_i$) are positive definite and positive semi-definite matrices, respectively. For simplicity and without loss of generality, assume that Q and R are symmetric. It is known that if (A, B) is stabilizable, then the performance index (2.3) is minimized by using the centralized state feedback law:

$$u(t) = -Kx(t) \tag{2.4}$$

where the gain matrix

$$K := \begin{bmatrix} k_{11} & \dots & k_{1\nu} \\ \vdots & \ddots & \vdots \\ k_{\nu 1} & \dots & k_{\nu\nu} \end{bmatrix}, \quad k_{ij} \in \Re^{m_i \times n_j}, \quad i,j \in \bar{\nu} \tag{2.5}$$

is derived from the solution of the Riccati equation [25].

Define the $\nu \times \nu$ block matrix $\Phi = sI - A + BK$ as

$$\Phi := \begin{bmatrix} \Phi_{11} & \dots & \Phi_{1\nu} \\ \vdots & \ddots & \vdots \\ \Phi_{\nu 1} & \dots & \Phi_{\nu\nu} \end{bmatrix}, \quad \Phi_{ij} \in \Re^{n_i \times n_j}, \quad i,j \in \bar{\nu} \tag{2.6}$$

and for any $i \in \bar{\nu}$, define:

$$\begin{aligned} M_{1_i}(s) &:= \begin{bmatrix} \Phi_{11} & \dots & \Phi_{1(i-1)} \\ \vdots & \ddots & \vdots \\ \Phi_{(i-1)1} & \dots & \Phi_{(i-1)(i-1)} \end{bmatrix}, \quad M_{2_i} := \begin{bmatrix} \Phi_{1(i+1)} & \dots & \Phi_{1\nu} \\ \vdots & \ddots & \vdots \\ \Phi_{(i-1)(i+1)} & \dots & \Phi_{(i-1)\nu} \end{bmatrix}, \\ M_{3_i} &:= \begin{bmatrix} \Phi_{(i+1)1} & \dots & \Phi_{(i+1)(i-1)} \\ \vdots & \ddots & \vdots \\ \Phi_{\nu 1} & \dots & \Phi_{\nu(i-1)} \end{bmatrix}, \quad M_{4_i}(s) := \begin{bmatrix} \Phi_{(i+1)(i+1)} & \dots & \Phi_{(i+1)\nu} \\ \vdots & \ddots & \vdots \\ \Phi_{\nu(i+1)} & \dots & \Phi_{\nu\nu} \end{bmatrix}. \end{aligned} \tag{2.7}$$

It is to be noted that the entries of the matrices $M_{1_i}(s)$ and $M_{4_i}(s)$ are functions of s, but the entries of the two other matrices are constant and independent of s. Consider now the following ν local controllers for the system (2.1):

$$
\begin{aligned}
U_i(s) = & \begin{bmatrix} k_{i1} & \dots & k_{i(i-1)} & k_{i(i+1)} & \dots & k_{iv} \end{bmatrix} \begin{bmatrix} M_{1_i}(s) & M_{2_i} \\ M_{3_i} & M_{4_i}(s) \end{bmatrix}^{-1} \\
& \times \begin{bmatrix} B_1 k_{1i} \\ \vdots \\ B_{(i-1)} k_{(i-1)i} \\ -A_{(i+1)i} + B_{(i+1)} k_{(i+1)i} \\ \vdots \\ -A_{vi} + B_v k_{vi} \end{bmatrix} X_i(s) \\
& - \begin{bmatrix} k_{i1} & \dots & k_{i(i-1)} & k_{i(i+1)} & \dots & k_{iv} \end{bmatrix} \begin{bmatrix} M_{1_i}(s) & M_{2_i} \\ M_{3_i} & M_{4_i}(s) \end{bmatrix}^{-1} \\
& \times \begin{bmatrix} x_0^{1,i} \\ \vdots \\ x_0^{i-1,i} \\ x_0^{i+1,i} \\ \vdots \\ x_0^{v,i} \end{bmatrix} - k_{ii} X_i(s) \ , \quad i \in \bar{v}
\end{aligned}
\tag{2.8}
$$

Theorem 1 *By choosing $x_0^{j,i} = x_j(0)$, $i, j \in \bar{v}$, $i \neq j$ in (2.8), the resultant decentralized control law will be equivalent to the optimal centralized controller (2.4).*

Proof Substitute (2.2), (2.4), and (2.5) into (2.1), take the Laplace transform of the resultant matrix equation, and eliminate its ith row. Rearrange the equation to obtain a relation between

$$
\begin{bmatrix} X_1(s)^T & X_2(s)^T & \cdots & X_{i-1}(s)^T & X_{i+1}(s)^T & \cdots & X_v(s)^T \end{bmatrix}^T \tag{2.9}
$$

and

$$
\begin{bmatrix} x_1(0)^T & \cdots & x_{i-1}(0)^T & x_{i+1}(0)^T & \cdots & x_v(0)^T \end{bmatrix}^T \tag{2.10}
$$

The proof follows immediately by substituting the resultant relation into the ith block row of the equation (2.4) in the Laplace domain. ■

Note that $U_i(s)$ in (2.8) is expressed in terms of the corresponding local information $X_i(s)$ and some constant values $x_0^{j,i}$, $j = 1,...,i-1,i+1,...,\nu$, but the parameters of the overall system A_{ji}, B_i, $i, j \in \overline{\nu}$, $i \geq j$, are assumed to be known by each subsystem. This assumption, however, is relaxed in Section 2.6. Note also that the control law given by (2.8) is time-invariant and incrementally linear due to the constants $x_0^{i,j}$, $i, j \in \bar{\nu}$, $i \neq j$.

Since the control law given by (2.8) depends on the constant values $x_0^{j,i}$, $i, j \in \overline{\nu}$, $j \neq i$, it is very important to check the stability of the system with the resultant decentralized controller.

2.4 Stability analysis via graph decomposition

It is desired now to find some conditions for the stability of the system (2.1) when the local controllers (2.8) are applied to the corresponding subsystems. The following definition will be used in Theorem 2.

Definition 1 *Consider the system $\mathscr{S}$ given by (2.1). The modified system $\mathbf{S}^i$, $i \in \{2,\ldots,\nu\}$, is defined to be a system obtained by removing all interconnections going to the ith subsystem in $\mathscr{S}$. The state equation of the modified system $\mathbf{S}^i$ is as follows:*

$$\dot{x} = \tilde{A}^i x + Bu$$

where $\tilde{A}^i$ is derived from A by replacing the first $i-1$ block entries of its ith block row with zeros.

Theorem 2 *Consider the system $\mathscr{S}$ given by (2.1). Assume that the ν local controllers given by (2.8) are applied to the corresponding subsystems. A sufficient and* almost always *necessary condition for stability of the resultant decentralized closed-loop system, regardless of the constant values $x_0^{j,i}$, $i, j \in \bar{\nu}$, $i \neq j$, is that all modified systems $\mathbf{S}^i$, $i = 2,\ldots,\nu$, are stable under the centralized state feedback law (2.4).*

Remark 2 *"Almost always necessary" in Theorem 2 means that for the given matrices A and B, the set of stabilizing gains K for which the stability of $\mathbf{S}^i$, $i = 2,\ldots,\nu$ under the centralized state feedback law (2.4) is violated but the proposed decentralized closed-loop system is still stable, is either an empty set or a hypersurface in the parameter space of K [26].*

Proof *Proof of sufficiency:* Suppose that the centralized LTI control system obtained by applying the state feedback law $u(t) = -Kx(t)$ to the modified system $\mathbf{S}^i$ is stable for all $i \in \{2, ..., \nu\}$. It will be proved by using strong induction that the states of the decentralized control system with the ν local controllers (2.8) are bounded.

Basis of induction $(i = 1)$: It is desired to show that the state of the first subsystem is bounded. However, the proof is omitted due to its similarity to the proof of the *induction step*, which will follow.

Induction hypothesis: Suppose that the state of the ith subsystem is bounded for $i = 1, 2, ..., m-1$.

Induction step: It is required to prove that the state of the mth subsystem is bounded. To simplify the formulation, define the following matrices ($m \in \bar{\nu}$):

$$Y_{1_m} := \begin{bmatrix} B_m k_{m1} & \dots & B_m k_{m(m-1)} \end{bmatrix}, \quad Y_{2_m} := \begin{bmatrix} B_m k_{m(m+1)} & \dots & B_m k_{m\nu} \end{bmatrix}$$
$$Z_{1_m} := \begin{bmatrix} B_1 k_{1m} \\ \vdots \\ B_{(m-1)} k_{(m-1)m} \end{bmatrix}, \quad Z_{2_m} := \begin{bmatrix} -A_{(m+1)m} + B_{(m+1)} k_{(m+1)m} \\ \vdots \\ -A_{\nu m} + B_\nu k_{\nu m} \end{bmatrix}$$
$$x_0^m := \begin{bmatrix} x_0^{1,m} \\ \vdots \\ x_0^{m-1,m} \\ x_0^{m+1,m} \\ \vdots \\ x_0^{\nu,m} \end{bmatrix}, \quad H_m(s) := [sI - A_{mm} + B_m k_{mm}] \tag{2.11}$$

Now, by using equations (2.8) and (2.1) (with the matrices A and B given by (2.2a)), the following can be concluded:

$$\begin{aligned} sX_m(s) &= A_{m1}X_1(s) + A_{m2}X_2(s) + \dots + A_{mm}X_m(s) - B_m k_{mm} X_m(s) \\ &+ \begin{bmatrix} Y_{1_m} & Y_{2_m} \end{bmatrix} \begin{bmatrix} M_{1_m}(s) & M_{2_m} \\ M_{3_m} & M_{4_m}(s) \end{bmatrix}^{-1} \begin{bmatrix} Z_{1_m} \\ Z_{2_m} \end{bmatrix} X_m(s) + x_m(0) \\ &- \begin{bmatrix} Y_{1_m} & Y_{2_m} \end{bmatrix} \begin{bmatrix} M_{1_m}(s) & M_{2_m} \\ M_{3_m} & M_{4_m}(s) \end{bmatrix}^{-1} x_0^m \end{aligned} \tag{2.12}$$

Based on the induction assumption, $x_j(t)$'s are bounded for $j = 1, 2, ..., m-1$, and consequently they can be considered as exponentially decaying disturbances for the mth subsystem. Hence, they do not influence the stability of the mth subsystem. Define the homogenous solution $x_{hm}(t)$ to be the part of the solution for $x_m(t)$ which corresponds to $x_1(t) = \cdots = x_{m-1}(t) = 0$. This solution satisfies the following equation:

$$
\begin{aligned}
sX_{hm}(s) = A_{mm}X_{hm}(s) - B_m k_{mm} X_{hm}(s) - \begin{bmatrix} Y_{1_m} & Y_{2_m} \end{bmatrix} \begin{bmatrix} M_{1_m}(s) & M_{2_m} \\ M_{3_m} & M_{4_m}(s) \end{bmatrix}^{-1} x_0^m \\
+ \begin{bmatrix} Y_{1_m} & Y_{2_m} \end{bmatrix} \begin{bmatrix} M_{1_m}(s) & M_{2_m} \\ M_{3_m} & M_{4_m}(s) \end{bmatrix}^{-1} \begin{bmatrix} Z_{1_m} \\ Z_{2_m} \end{bmatrix} X_{hm}(s) + x_m(0)
\end{aligned}
\tag{2.13}
$$

or equivalently

$$
\begin{aligned}
\left((sI - A_{mm} + B_m k_{mm}) - \begin{bmatrix} Y_{1_m} & Y_{2_m} \end{bmatrix} \begin{bmatrix} M_{1_m}(s) & M_{2_m} \\ M_{3_m} & M_{4_m}(s) \end{bmatrix}^{-1} \begin{bmatrix} Z_{1_m} \\ Z_{2_m} \end{bmatrix} \right) X_{hm}(s) = \\
x_m(0) - \begin{bmatrix} Y_{1_m} & Y_{2_m} \end{bmatrix} \begin{bmatrix} M_{1_m}(s) & M_{2_m} \\ M_{3_m} & M_{4_m}(s) \end{bmatrix}^{-1} x_0^m
\end{aligned}
\tag{2.14}
$$

It can be concluded from (2.14) that $x_{hm}(t)$ can be expressed as:

$$
\sum_{i=1}^{l} \left(p_i(t) x_m(0) + q_i(t) x_0^m \right) e^{s_i t} \tag{2.15}
$$

where $s = s_i$, $i = 1, 2, ..., l$, are the roots of the following equation

$$
\begin{aligned}
det \left(H_m(s) - \begin{bmatrix} Y_{1_m} & Y_{2_m} \end{bmatrix} \begin{bmatrix} M_{1_m}(s) & M_{2_m} \\ M_{3_m} & M_{4_m}(s) \end{bmatrix}^{-1} \begin{bmatrix} Z_{1_m} \\ Z_{2_m} \end{bmatrix} \right) \\
\times\, det \begin{bmatrix} M_{1_m}(s) & M_{2_m} \\ M_{3_m} & M_{4_m}(s) \end{bmatrix} = 0
\end{aligned}
\tag{2.16}
$$

and also $p_i(t)$ and $q_i(t)$, $i = 1, 2, ..., l$ are matrices with polynomial entries of degree less than or equal to the multiplicity of $s = s_i$ as the root of the above equation, minus one. On the other hand, it can be shown that:

$$
\begin{aligned}
det \begin{bmatrix} L_1 & L_2 & L_3 \\ L_4 & L_5 & L_6 \\ L_7 & L_8 & L_9 \end{bmatrix} = det \begin{bmatrix} L_1 & L_3 \\ L_7 & L_9 \end{bmatrix} \\
\times\, det \left(L_5 - \begin{bmatrix} L_4 & L_6 \end{bmatrix} \begin{bmatrix} L_1 & L_3 \\ L_7 & L_9 \end{bmatrix}^{-1} \begin{bmatrix} L_2 \\ L_8 \end{bmatrix} \right)
\end{aligned}
$$

where L_1, L_5, and L_9 are square matrices and L_1, L_3, L_7, and L_9 are matrices with the property that $\begin{bmatrix} L_1 & L_3 \\ L_7 & L_9 \end{bmatrix}$ is nonsingular. Thus, the equation (2.16) can be simplified as follows:

$$det \begin{bmatrix} M_{1_m}(s) & Z_{1_m} & M_{2_m} \\ Y_{1_m} & H_m(s) & Y_{2_m} \\ M_{3_m} & Z_{2_m} & M_{4_m}(s) \end{bmatrix} = 0 \tag{2.17}$$

By substituting the entries of the above matrix from (2.7) and (2.11), it can be rewritten in the following simplified form:

$$det(sI - \tilde{A}^m + BK) = 0 \tag{2.18}$$

On the other hand, the modes of the closed-loop system $\mathbf{S}^m$ under the feedback law (2.4) can be obtained from (2.18). Since it has been assumed that this closed-loop system is stable, all complex numbers $s_1, ..., s_l$ will be in the open left-half s-plane. As a result, the state of the mth subsystem is bounded.

Proof of necessity for almost all K's: Suppose that some of the modified systems $\mathbf{S}^2, \mathbf{S}^3, ..., \mathbf{S}^v$ are not stabilized by the feedback law (2.4). It is desired to show that the system $\mathscr{S}$ under the proposed local controllers (2.8) is *almost always* unstable. Let the first modified system which is unstable under the feedback law (2.4) be denoted by $\mathbf{S}^m$, i.e. all of the systems $\mathbf{S}^2, \mathbf{S}^3, ..., \mathbf{S}^{m-1}$ are stabilized by (2.4). Using the first $m-1$ steps of the induction introduced in the proof of sufficiency, it can be concluded that the states of the subsystems $1, 2, ..., m-1$ of the system $\mathscr{S}$ under the proposed local controllers (2.8) are bounded. Now, if the induction continues one more step, it can be concluded that since $x_j(t)$ is bounded for $j = 1, 2, ..., m-1$, there exists a particular solution for $x_m(t)$ which approaches zero as time goes to infinity, and the homogenous part of the solution for $x_m(t)$ (denoted by $x_{hm}(t)$, which corresponds to $x_1(t) = \cdots = x_{m-1}(t) = 0$) satisfies the equation (2.13), or equivalently (2.14). Choose any arbitrary unstable mode of the modified system $\mathbf{S}^m$ under the feedback law (2.4), and denote it with $s = \sigma^m$. This mode must satisfy (2.17) or equivalently (2.16). As mentioned in the proof of sufficiency, $x_{hm}(t)$ can be expressed as $\sum_{i=1}^{l} \left(p_i(t)x_m(0) + q_i(t)x_0^m\right) e^{s_i t}$, where $s = s_i$, $i = 1, 2, ..., l$, are the roots of the equation (2.16). However, it is required to determine whether or not $s = \sigma^m$ satisfying (2.16) appears among $s = s_i$, $i = 1, 2, ..., l$. It can be easily verified that $\sigma^m \neq s_i$, for all $i = 1, 2, ..., l$ iff both of the following conditions hold.

- $s=\sigma^m$ is a root of the following equation:

$$det\begin{bmatrix} M_{1_m}(s) & M_{2_m} \\ M_{3_m} & M_{4_m}(s) \end{bmatrix}=0 \tag{2.19}$$

  Note that if the above equation is not satisfied for $s=\sigma^m$, then

$$det\left(H_m(s)-\begin{bmatrix} Y_{1_m} & Y_{2_m} \end{bmatrix}\begin{bmatrix} M_{1_m}(s) & M_{2_m} \\ M_{3_m} & M_{4_m}(s) \end{bmatrix}^{-1}\begin{bmatrix} Z_{1_m} \\ Z_{2_m} \end{bmatrix}\right)=0$$

  for $s=\sigma^m$. This will generate a term $p_i(t)x_m(0)e^{\sigma^m t}$ in $x_{hm}(t)$ that makes $x_m(t)$ go to infinity as time increases. Since the matrix in the left side of (2.19) has been derived from $sI-A+BK$ by eliminating its mth block row and m^{th} block column, this requirement is equivalent to the following statement:

  The modified system $\mathbf{S}^m$ has an unstable mode $s=\sigma^m$ under the feedback law (2.4), and that mode is also the unstable mode of the system $\mathscr{S}$ under the feedback law (2.4) after isolating its mth subsystem (eliminating all of the inputs, outputs, and interconnections of its mth subsystem).

- The mode $s=\sigma^m$ is cancelled out in the following expression:

$$\begin{bmatrix} Y_{1_m} & Y_{2_m} \end{bmatrix}\begin{bmatrix} M_{1_m}(s) & M_{2_m} \\ M_{3_m} & M_{4_m}(s) \end{bmatrix}^{-1}$$

  This means that $s=\sigma^m$ does not appear in any of the denominators of the entries of the above matrix. Let the matrix obtained from $A-BK$ by eliminating its mth block row and m^{th} block column be denoted by Φ^m. It is easy to verify that this condition is equivalent to the following statement:

  The mode $s=\sigma^m$ is an unobservable mode of the pair $\left(\begin{bmatrix} Y_{1_m} & Y_{2_m} \end{bmatrix},\Phi^m\right)$.

Apparently, for the given matrices A and B, the set of stabilizing gains K for which both of the above conditions hold is either an empty set or a hypersurface in the parameter space of K (for definition of a hyperssurface and some similar examples see [26]). If the stabilizing gains K located on a hypersurface are neglected, $s=\sigma^m$ appears among $s=s_i$, $i=1,2,...,l$, which makes $x_{hm}(t)$ go to infinity, as t increases. This yields the instability of the decentralized closed-loop control systems. ■

Theorem 2 states that the stability of the interconnected system given by (2.1) under the proposed decentralized control law is *almost always* equivalent to the stability of a set of $v-1$ centralized LTI control systems, which can be easily verified from the location of the corresponding eigenvalues.

2.5 Robust stability analysis

Since the decentralized control law (2.8) has been obtained based on the nominal parameters of the system $\mathscr{S}$, it may become unstable once the proposed control law is applied to the perturbed system. Thus, the robust stability of the controller with respect to uncertainties in the original system is an important issue which will be addressed in this section.

Suppose that the decentralized control law (2.8), which is computed in terms of the nominal parameters A and B of the system $\mathscr{S}$, is applied to the system $\bar{\mathscr{S}}$, which is the perturbed version of $\mathscr{S}$ described as follows:

$$\dot{x} = \bar{A}x + \bar{B}u \tag{2.20}$$

where

$$\bar{A} := \begin{bmatrix} \bar{A}_{11} & 0 & \dots & 0 \\ \bar{A}_{21} & \bar{A}_{22} & \dots & 0 \\ \vdots & \vdots & \ddots & \vdots \\ \bar{A}_{\nu 1} & \bar{A}_{\nu 2} & \dots & \bar{A}_{\nu\nu} \end{bmatrix}, \quad \bar{B} := \begin{bmatrix} \bar{B}_1 & \dots & 0 \\ \vdots & \ddots & \vdots \\ 0 & \dots & \bar{B}_\nu \end{bmatrix}$$

It is to be noted that the perturbed matrices $\bar{A}$ and $\bar{B}$ are also block lower triangular and block diagonal, respectively. In other words, it is assumed that parameter variations will not generate new interconnections, i.e. the structural graph of the perturbed system will also be acyclic.

Definition 2 *The perturbed modified system $\bar{\mathbf{S}}^i, i \in \bar{\nu}$, is defined by:*

$$\dot{x} = \bar{A}^i x + \bar{B}^i u$$

where the matrix $\bar{A}^i$ is the same as A, except for its $i-1$ block entries $A_{i1},\dots,A_{i(i-1)}$, which are replaced by zeros, and its A_{ii} block entry which is replaced by $\bar{A}_{ii}$. Also, the matrix $\bar{B}^i$ is the same as B, except for its (i,i) block entry B_i, which is replaced by $\bar{B}_i$. $\bar{\mathbf{S}}^i$ is, in fact, obtained by modifying $\mathscr{S}$ as follows:

- *All interconnections going to the ith subsystem are removed.*
- *The nominal parameters (A_{ii}, B_i) of the ith subsystem are replaced by the perturbed parameters $(\bar{A}_{ii}, \bar{B}_i)$.*

Theorem 3 *Consider the system $\bar{\mathscr{S}}$ given by (2.20), and assume that the ν local controllers given by (2.8) are applied to the corresponding subsystems. A sufficient and* almost always *necessary condition for stability of the resultant decentralized closed-loop system, regardless of the constant values $x_0^{j,i}$, $i, j \in \bar{\nu}$, $i \neq j$, is that all perturbed modified systems $\bar{\mathbf{S}}^i$, $i = 1, \ldots, \nu$, are stable under the centralized state feedback law (2.4).*

Proof The proof is omitted due to its similarity to the proof of Theorem 2. ∎

Remark 3 *Since none of the perturbed parameters $\bar{A}_{ij}$, $i, j \in \bar{\nu}$, $i \neq j$, appear in the perturbed modified systems $\bar{\mathbf{S}}^1, \bar{\mathbf{S}}^2, \ldots, \bar{\mathbf{S}}^\nu$, the robust stability of the proposed decentralized feedback law is independent of the perturbation of the interconnection parameters (note that this statement is valid for any decentralized control law designed by any arbitrary approach, which is applied to an acyclic interconnected system. In other words, the controller need not be optimal).*

Define the perturbation matrix as the perturbed matrix minus the original matrix. The perturbed and perturbation matrices for a matrix M are denoted by $\bar{M}$ and ΔM, respectively. Suppose that the decentralized feedback law (2.8) is designed in terms of the nominal matrices A and B, and then applied to the perturbed system $\bar{\mathscr{S}}$ with the state-space matrices $\bar{A}$ and $\bar{B}$. The objective is to find the allowable perturbation matrices $\Delta A = \bar{A} - A$ and $\Delta B = \bar{B} - B$, for which the decentralized closed-loop system will still remain stable. In Theorem 3, a sufficient condition to achieve this objective is presented, which is *almost always* necessary. Robustness analysis with respect to the perturbation in the parameters of the system can then be summarized as follows:

- For decentralized case, the location of the eigenvalues of the ν matrices $\bar{A}^1 - \bar{B}^1 K, \bar{A}^2 - \bar{B}^2 K, \ldots,$ $\bar{A}^\nu - \bar{B}^\nu K$ should be checked.

- For centralized case, the location of the eigenvalues of the matrix $\bar{A} - \bar{B}K$ should be checked.

Robustness analysis in both classes addresses the following problem, in general:
Consider a Hurtiwz matrix M, and assume that its entries are subject to perturbations. It is desired to know how much sensitive the eigenvalues of M are to the variation of its entries. More specifically, it is desired to find out how much the matrix M can be perturbed so that the resultant matrix is still Hurtiwz.

This problem has been addressed in the literature using different mathematical approaches [28], [29], [30]. Sensitivity of the eigenvalues to the variation of its entries depends, in general, on several factors such as the norm of the perturbation matrix, structure of the matrix (represented by condition number or eigenvalue condition number [29]), and repetition or distinction of the eigenvalues.

Theorem 4 *The bound on the Frobenius norm of the perturbation matrix corresponding to each of the matrices $\bar{A}^i - \bar{B}^i K$, $i = 1,2,\dots,\nu$ in the decentralized case is less than or equal to that of the perturbation matrix corresponding to $\bar{A} - \bar{B}K$ in the centralized case.*

Proof The following relation holds for the decentralized case:

$$||\Delta(\bar{A}^i - \bar{B}^i K)||_F = ||(\bar{A}^i - \bar{B}^i K) - (\tilde{A}^i - BK)||_F = \sqrt{\sum_{j=1, j\neq i}^{\nu} ||\Delta B_i k_{ij}||_F^2 + ||\Delta B_i k_{ii} + \Delta A_{ii}||_F^2}$$

This results in:

$$||\Delta(\bar{A}^i - \bar{B}^i K)||_F \leq \sqrt{\Gamma_{\text{dec}_i}}, \qquad i \in \bar{\nu} \tag{2.21}$$

where

$$\Gamma_{\text{dec}_i} := \sum_{j=1}^{\nu} ||\Delta B_i k_{ij}||_F^2 + ||\Delta A_{ii}||_F^2, \qquad i \in \bar{\nu} \tag{2.22}$$

For the centralized case, on the other hand, one can write

$$\begin{aligned} ||\Delta(\bar{A} - \bar{B}K)||_F &= ||(\bar{A} - \bar{B}K) - (A - BK)||_F \\ &= \sqrt{\sum_{i=1}^{\nu}\sum_{j=1}^{i} ||\Delta B_i k_{ij} + \Delta A_{ij}||_F^2 + \sum_{i=1}^{\nu}\sum_{j=i+1}^{\nu} ||\Delta B_i k_{ij}||_F^2} \end{aligned}$$

Thus,

$$||\Delta(\bar{A} - \bar{B}K)||_F \leq \sqrt{\Gamma_{\text{cen}}} \tag{2.23}$$

where

$$\Gamma_{\text{cen}} := \sum_{i=1}^{\nu}\sum_{j=1}^{\nu} ||\Delta B_i k_{ij}||_F^2 + \sum_{i=1}^{\nu}\sum_{j=1}^{i} ||\Delta A_{ij}||_F^2 \tag{2.24}$$

It is apparent from (2.22) and (2.24), that

$$\Gamma_{\text{dec}_1} + \Gamma_{\text{dec}_2} + \cdots + \Gamma_{\text{dec}_\nu} \leq \Gamma_{\text{cen}} \tag{2.25}$$

Therefore,

$$\sqrt{\Gamma_{\text{dec}_i}} \leq \sqrt{\Gamma_{\text{cen}}}\,, \quad i = 1, 2, ..., \nu \tag{2.26}$$

The proof follows immediately from (2.21), (2.23) and (2.26). It is to be noted that the inequality (2.26) obtained above, is more conservative than (2.25), obtained in the preceding step. ■

According to Theorem 4, there are ν perturbed matrices in the decentralized case, and the bound on the Frobenius norm of the perturbation matrix for each of them is less than or equal to the bound on the Frobenius norm of the corresponding perturbation matrix in the centralized case. Therefore, it can be concluded from the above discussion and Remark 3, that the proposed decentralized controller is expected to perform better than the centralized counterpart in terms of robust stability with respect to the parameter variations of the system. This result can also be deduced intuitively, because for any subsystems i and j ($i > j$):

- In the centralized case, any perturbation in subsystem i will influence the state of subsystem j through the feedback and can cause the instability of the closed-loop system.
- In the decentralized case, no perturbation in subsystem i can influence the state of subsystem j through the feedback or through the interconnections, due to the particular structure of the system (i.e., lower-triangular structure of A and diagonal structure of B).

2.6 Non-identical local beliefs about the system model

In practice, different local controllers may assume different models for the overall system. It is desired now to find some results similar to the ones presented in Theorem 3, under the above condition.

Suppose that control agent l assumes the matrices $\hat{A}^l$ and $\hat{B}^l$ instead of the matrices A and B in the state-space representation (2.1) of the system $\mathscr{S}$. Denote the (i, j) block entry of $\hat{A}^l$ with $\hat{A}^l_{ij} \in \Re^{n_i \times n_j}$, for any $i, j \in \bar{\nu}$, $i \geq j$, and the (i, i) block entry of $\hat{B}^l$ with $\hat{B}^l_i \in \Re^{n_i \times m_i}$ for any $i \in \bar{\nu}$. Now, for any $l \in \bar{\nu}$, replace A and B in the equation (2.1) with $\hat{A}^l$ and $\hat{B}^l$, respectively. Then solve the corresponding LQR problem for the above matrices, to obtain the optimal static gain $\hat{K}^l$, whose (i, j) block entry is denoted by $\hat{k}^l_{ij}$, for all $i, j \in \bar{\nu}$. Define the matrices $\hat{M}^l_{1_i}(s), \hat{M}^l_{2_i}, \hat{M}^l_{3_i}$ and $\hat{M}^l_{4_i}(s)$ similarly to the matrices in (2.7), by replacing $\Phi = sI - A + BK$ in (2.6) with $\hat{\Phi}^l := sI - \hat{A}^l + \hat{B}^l\hat{K}^l$. Therefore, the lth local control law, in this case, is given by ($l \in \bar{\nu}$):

$$
\begin{aligned}
U_l(s) = &\begin{bmatrix} \hat{k}^l_{l1} & \dots & \hat{k}^l_{l(l-1)} & \hat{k}^l_{l(l+1)} & \dots & \hat{k}_{l\nu} \end{bmatrix} \begin{bmatrix} \hat{M}^l_{1_l}(s) & \hat{M}^l_{2_l} \\ \hat{M}^l_{3_l} & \hat{M}^l_{4_l}(s) \end{bmatrix}^{-1} \\
&\times \begin{bmatrix} \hat{B}^l_1 \hat{k}^l_{1l} \\ \vdots \\ \hat{B}^l_{(l-1)} \hat{k}^l_{(l-1)l} \\ -\hat{A}^l_{(l+1)l} + \hat{B}^l_{(l+1)} \hat{k}^l_{(l+1)l} \\ \vdots \\ -\hat{A}^l_{\nu l} + \hat{B}^l_\nu \hat{k}^l_{\nu l} \end{bmatrix} X_l(s) \\
&- \begin{bmatrix} \hat{k}^l_{l1} & \dots & \hat{k}^l_{l(l-1)} & \hat{k}^l_{l(l+1)} & \dots & \hat{k}^l_{l\nu} \end{bmatrix} \begin{bmatrix} \hat{M}^l_{1_l}(s) & \hat{M}^l_{2_l} \\ \hat{M}^l_{3_l} & \hat{M}^l_{4_l}(s) \end{bmatrix}^{-1} \\
&\times \begin{bmatrix} x_0^{1,l} \\ \vdots \\ x_0^{l-1,l} \\ x_0^{l+1,l} \\ \vdots \\ x_0^{\nu,l} \end{bmatrix} - \hat{k}^l_{ll} X_l(s)
\end{aligned}
\tag{2.27}
$$

Definition 3 *The uncertain model $\hat{\mathbf{S}}^l$, $l \in \bar{\nu}$, is defined by:*

$$\dot{x}(t) = A^l x(t) + B^l u(t)$$

where the matrix A^l is the same as $\hat{A}^l$, except for its $l-1$ block entries $\hat{A}^l_{l1}, \dots, \hat{A}^l_{l(l-1)}$, which are replaced by zeros, and its $\hat{A}^l_{ll}$ block entry, which is replaced by $\bar{A}_{ll}$. Also, the matrix B^l is the same as $\hat{B}^l$, except for its (l,l) block entry $\hat{B}^l_l$, which is replaced by $\bar{B}_l$.

Corollary 1 *Consider the system $\bar{\mathscr{S}}$ given by (2.20). Assume that the ν local controllers given by (2.27) are applied to the corresponding subsystems. A sufficient and* almost always *necessary condition for stability of the resultant decentralized closed-loop system, regardless of the constant values*

$x_0^{j,i}$, $i, j \in \bar{\mathrm{v}}$, $i \neq j$, *is that the uncertain system* $\hat{\mathbf{S}}^i$ *is stable under the centralized state feedback law* $u(t) = -\hat{K}^i x(t)$, *for all* $i \in \bar{\mathrm{v}}$.

Proof The proof is omitted due to its similarity to the proof of Theorem 2. ■

2.7 Centralized and decentralized performance comparison

So far, a decentralized control law has been proposed for a class of stabilizable LTI systems with the property that if the modeling parameters and the initial state of each subsystem are available in all other subsystems, then the proposed controller will be equivalent to the optimal centralized controller. It is to be noted that the equalities $\hat{A}^l = A$, $\hat{B}^l = B$, $l \in \bar{\mathrm{v}}$, will hereafter be assumed to simplify the presentation of the properties of the decentralized control proposed in this chapter. Note that the results presented under this assumption, can simply be extended to the general case. It has also been shown that if the conditions of Theorem 2 are satisfied, then by using any arbitrary constant values instead of the initial states of other subsystems form each subsystem's view, the resultant decentralized closed-loop system will remain stable, which implies that the deviation ΔJ of the resultant quadratic performance index (2.3) from the optimal performance index corresponding to the centralized LQR controller remains finite. The following definitions are used to find ΔJ.

Definition 4 *Define* $\Delta_i x_j(0)$, $i, j \in \overline{\mathrm{v}}$, $i \neq j$, *as the difference between the initial state of the* j^{th} *subsystem* $x_j(0)$ *and* $x_0^{j,i}$. *Throughout the remainder of the chapter, this difference will be referred to as the prediction error of the initial state.*

Due to the prediction errors defined above, there will be a deviation in the state $x_i(t)$ and control input $u_i(t)$, $i \in \overline{\mathrm{v}}$, of the resultant decentralized control system compared to those of the centralized counterpart. Denote the state and the control input deviations with $\Delta x_i(t)$ and $\Delta u_i(t)$, respectively.

Definition 5 *The matrices Δx_0 and $\Delta_m x(0)$, $m \in \bar{\nu}$, are defined as follows:*

$$\Delta x_0 = \begin{bmatrix} \Delta_1 x(0) \\ \Delta_2 x(0) \\ \vdots \\ \Delta_n x(0) \end{bmatrix} , \quad \Delta_m x(0) = \begin{bmatrix} \Delta_m x_1(0) \\ \vdots \\ \Delta_m x_{i-1}(0) \\ \Delta_m x_{i+1}(0) \\ \vdots \\ \Delta_m x_\nu(0) \end{bmatrix} , \quad m = 1,2,\ldots,\nu \tag{2.28}$$

The following algorithm is presented to find ΔJ in terms of the prediction errors $\Delta_i x_j(0)$, $i, j \in \overline{\nu}$, $i \neq j$.

Algorithm 1

1) *Find $\Delta X_1(s)$ in terms of $\Delta_1 x(0)$ by using equation (2.12) (for $m = 1$), which can be expressed as $\Delta X_1(s) = F_{11}(s)\Delta_1 x(0)$. Substitute $\Delta X_1(s)$ into equation (2.8) for $i = 1$ to obtain $\Delta U_1(s)$ only in terms of $\Delta_1 x(0)$, i.e. $\Delta U_1(s) = G_{11}(s)\Delta_1 x(0)$.*

⋮

m) *Assume that $\Delta X_i(s)$ and $\Delta U_i(s)$ have been computed for $i = 1,2,\ldots,m-1$ in terms of prediction errors in the previous steps of the algorithm. Now, for $i = m$, substitute the expressions obtained for $\Delta X_1(s), \Delta X_2(s), \ldots, \Delta X_{m-1}(s)$ into equation (2.12) to find an equation relating $\Delta X_m(s)$ to the prediction errors. Let this equation be represented by $\Delta X_m(s) = F_{m1}(s)\Delta_1 x(0) + F_{m2}(s)\Delta_2 x(0) + \cdots + F_{mm}(s)\Delta_m x(0)$. By substituting this expression into (2.8) for $i = m$, $\Delta U_m(s)$ will be found in terms of the prediction errors, i.e. $\Delta U_m(s) = G_{m1}(s)\Delta_1 x(0) + G_{m2}(s)\Delta_2 x(0) + \cdots + G_{mm}(s)\Delta_m x(0)$.*

⋮

The algorithm continues up to step ν. It is obvious from the expressions in step m of Algorithm 1, that the deviation in the state of each subsystem depends not only on its own prediction errors, but also on the prediction errors of all previous subsystems due to the interconnections. The results obtained from the algorithm can be written in the matrix form as follows:

$$\Delta X(s) = F(s)\Delta x_0 \quad , \quad \Delta U(s) = G(s)\Delta x_0$$

where

$$F(s) = \begin{bmatrix} F_{11}(s) & 0 & 0 & \dots & 0 \\ F_{21}(s) & F_{22}(s) & 0 & \dots & 0 \\ \vdots & \vdots & \vdots & \ddots & \vdots \\ F_{v1}(s) & F_{v2}(s) & F_{v3}(s) & \dots & F_{vv}(s) \end{bmatrix}, \quad \Delta X(s) = \begin{bmatrix} \Delta X_1(s) \\ \vdots \\ \Delta X_v(s) \end{bmatrix},$$

$$G(s) = \begin{bmatrix} G_{11}(s) & 0 & 0 & \dots & 0 \\ G_{21}(s) & G_{22}(s) & 0 & \dots & 0 \\ \vdots & \vdots & \vdots & \ddots & \vdots \\ G_{v1}(s) & G_{v2}(s) & G_{v3}(s) & \dots & G_{vv}(s) \end{bmatrix}$$

Therefore, the deviation of the performance index due to the prediction errors can be obtained as follows:

$$\begin{aligned} \Delta J &= \int_0^\infty \left([x+\Delta x]^T Q[x+\Delta x] + [u+\Delta u]^T R[u+\Delta u]\right) dt - \int_0^\infty \left(x^T Qx + u^T Ru\right) dt \\ &= \int_0^\infty \left(x^T Q\Delta x + \Delta x^T Qx + \Delta x^T Q\Delta x + u^T R\Delta u + \Delta u^T Ru + \Delta u^T R\Delta u\right) dt \end{aligned} \tag{2.29}$$

It is to be noted that x and u are the state and the input of the centralized closed-loop system, and $x+\Delta x$ and $u+\Delta u$ are those of the decentralized closed-loop system. On the other hand, equations (2.1) and (2.4) yield:

$$X(s) = W(s)x(0) \quad , \quad U(s) = Z(s)x(0)$$

where

$$W(s) = (SI - A + BK)^{-1}, \quad Z(s) = -K(SI - A + BK)^{-1}$$

Suppose that $w(t), z(t), f(t)$ and $g(t)$ represent the time domain functions corresponding to $W(s), Z(s), F(s)$ and $G(s)$, respectively. Substituting these time functions into (2.29) results in:

$$\begin{aligned} \Delta J = &\int_0^\infty \left(x(0)^T w(t)^T Qf(t)\Delta x_0 + \Delta x_0^T f(t)^T Qw(t)x(0) + \Delta x_0^T f(t)^T Qf(t)\Delta x_0\right) dt \\ &+ \int_0^\infty \left(x(0)^T z(t)^T Rg(t)\Delta x_0 + \Delta x_0^T g(t)^T Rz(t)x(0) + \Delta x_0^T g(t)^T Rg(t)\Delta x_0\right) dt \end{aligned}$$

Due to the causality of the system, the arguments of both integrals in the above equation are zero for negative time. As a result, the interval for both integrals can be changed from $(0, +\infty)$ to $(-\infty, +\infty)$.

Hence, one can use the Parseval's formula to obtain:

$$
\begin{aligned}
\Delta J = &\frac{1}{2\pi}\int_{-\infty}^{\infty}\left(x(0)^T W(j\omega)^T QF(-j\omega)\Delta x_0 + \Delta x_0^T F(j\omega)^T QW(-j\omega)x(0)\right.\\
&\left.+\Delta x_0^T F(j\omega)^T QF(-j\omega)\Delta x_0\right) d\omega\\
&+\frac{1}{2\pi}\int_{-\infty}^{\infty}\left(x(0)^T Z(j\omega)^T RG(-j\omega)\Delta x_0 + \Delta x_0^T G(j\omega)^T RZ(-j\omega)x(0)\right.\\
&\left.+\Delta x_0^T G(j\omega)^T RG(-j\omega)\Delta x_0\right) d\omega
\end{aligned} \tag{2.30}
$$

Define the following matrices:

$$
\begin{aligned}
V_{12} &= \frac{1}{2\pi}\int_{-\infty}^{\infty}\left(W(j\omega)^T QF(-j\omega) + Z(j\omega)^T RG(-j\omega)\right) d\omega\\
V_{21} &= \frac{1}{2\pi}\int_{-\infty}^{\infty}\left(F(j\omega)^T QW(-j\omega) + G(j\omega)^T RZ(-j\omega)\right) d\omega\\
V_{22} &= \frac{1}{2\pi}\int_{-\infty}^{\infty}\left(F(j\omega)^T QF(-j\omega) + G(j\omega)^T RG(-j\omega)\right) d\omega
\end{aligned}
$$

It is to be noted that since R and Q are assumed to be symmetric matrices, V_{21} and V_{22} are equal to V_{12}^T and V_{22}^T, respectively. Thus, it can be concluded from (2.30) that:

$$
\begin{aligned}
\Delta J &= x(0)^T V_{12}\Delta x_0 + \Delta x_0^T V_{21} x(0) + \Delta x_0^T V_{22}\Delta x_0\\
&= \begin{bmatrix} x(0)^T & \Delta x_0^T \end{bmatrix}\begin{bmatrix} 0 & V_{12}\\ V_{21} & V_{22}\end{bmatrix}\begin{bmatrix} x(0)\\ \Delta x_0\end{bmatrix}\\
&= \begin{bmatrix} x(0)^T & \Delta x_0^T \end{bmatrix}\begin{bmatrix} 0 & V_{12}\\ V_{12}^T & V_{22}\end{bmatrix}\begin{bmatrix} x(0)\\ \Delta x_0\end{bmatrix}
\end{aligned} \tag{2.32}
$$

Proposition 1 *The performance deviation ΔJ can be written as:*

$$
\Delta J = \Delta x_0^T V_{22}\Delta x_0 \tag{2.33}
$$

Proof Consider an arbitrary $x(0)$, and assume that Δx_0 is a variable vector. Note that the entries of Δx_0 can take any values, because they represent prediction errors of the initial states. ΔJ given in (2.32) has the following properties:

- ΔJ is always nonnegative, because the centralized optimal performance index has the smallest value among all performance indices resulted by using any type of centralized or decentralized controller.

- Substituting $\Delta x_0 = 0$ in (2.32) yields $\Delta J = 0$.

- ΔJ is continuous with respect to each of the entries of the variable Δx_0, because ΔJ is quadratic.

It can be concluded from the above properties that $\Delta x_0 = 0$ is an extremum point for ΔJ. Thus, the partial derivative of ΔJ with respect to Δx_0 is equal to zero at $\Delta x_0 = 0$. Hence:

$$\left[x(0)^T V_{12} + \left(V_{12}^T x(0)\right)^T + \Delta x_0^T \left(V_{22}^T + V_{22}\right)\right]\Bigg|_{\Delta x_0 = 0} = 0$$

which results in $x(0)^T V_{12} = 0$. This implies that any arbitrary vector $x(0)$ is in the null space of V_{12}, or equivalently $V_{12} = 0$. The proof follows immediately from substituting $V_{12} = 0$ into (2.32), and noting that $V_{21} = V_{12}^T = 0$. ■

Remark 4 *Equation (2.33) states that for finding the performance deviation ΔJ, there is no need to obtain the time functions $w(t)$ and $z(t)$. In other words, only the functions $f(t)$ and $g(t)$ are required for performance evaluation. Furthermore, one can directly use the Laplace transforms $F(s)$ and $G(s)$, and substitute $s = \pm j\omega$ to obtain ΔJ through (2.31) and (2.33).*

Remark 5 *It can be concluded from (2.33), that the performance deviation ΔJ depends only on the prediction error of the initial state, not the initial state itself.*

Theorem 5 *To minimize the expected value of the performance index J, the constant value $x_0^{j,i}$ should be chosen equal to the expected value of the jth subsystem's initial state from the ith subsystem's view for any $i, j \in \bar{\mathbf{v}}$, $i \neq j$.*

Proof Consider an arbitrary initial state $x(0)$. It can be concluded from (2.28) and Definition 4 that Δx_0 can be written as $\hat{x}_0 - x_0$, where $\hat{x}_0$ is a vector whose entries are related to the constant values $x_0^{j,m}$. Also, x_0 is a vector whose entries are related to the initial states $x_m(0)$, $m \in \bar{\mathbf{v}}$. Consequently,

$$\begin{aligned} E\{\Delta J\} &= E\{\Delta x_0^T V_{22} \Delta x_0\} = E\{(\hat{x}_0^T - x_0^T) V_{22} (\hat{x}_0 - x_0)\} \\ &= \hat{x}_0^T V_{22} \hat{x}_0 - E\{x_0\}^T V_{22} \hat{x}_0 - \hat{x}_0^T V_{22} E\{x_0\} + E\{x_0\}^T V_{22} E\{x_0\} \end{aligned}$$

To minimize the above expression, take its partial derivative with respect to $\hat{x}_0$ and equate it to zero as follows:

$$\hat{x}_0^T (V_{22} + V_{22}^T) - E\{x_0\}^T V_{22} - (V_{22} E\{x_0\})^T = 0$$

which results in:

$$(\hat{x}_0^T - E\{x_0\}^T)(V_{22}+V_{22}^T) = 0 \tag{2.34}$$

Since the optimal control strategy is unique, ΔJ should be positive for any nonzero Δx_0. As a result, the matrix V_{22} in (2.33) is positive definite and consequently, the matrix $V_{22}+V_{22}^T$ is positive definite as well. Thus, the determinant of the matrix $V_{22}+V_{22}^T$ is nonzero, and so it can be concluded from (2.34) that $\hat{x}_0^T - E\{x_0\}^T = 0$, or equivalently $E\{\Delta x_0\} = E\{\hat{x}_0 - x_0\} = 0$. In other words, the expected value of any entry of Δx_0 should be zero. Thus, it can be deduced from (2.28) and Definition 4 that

$$E\{x_0^{j,m} - x_j(0)\} = 0, \quad j,m \in \bar{\nu},\ j \neq m$$

This relation states that the best choice for $x_0^{j,m}$ is equal to $E_m\{x_j(0)\}$, the expected value of the initial state of the jth subsystem from the mth subsystem's view. ■

Remark 6 *One can use Proposition 1 and Theorem 5 to obtain statistical results for the performance deviation ΔJ in terms of the expected values of the initial states of the subsystems. This can be achieved by using Chebyshev's inequality. This enables the designer to determine the maximum allowable standard deviation for Δx_0 to achieve a performance deviation within a prespecified region with a sufficiently high probability (e.g. 95%).*

Remark 7 *Suppose that the initial state of an acyclic interconnected system is a random variable whose mean $\bar{x}_0$ and covariance matrix are given. Consider a decentralized control law obtained by using the method in [16] (i.e., the second approach discussed in the introduction). For any given initial state $x(0)$, compute the quadratic performance index (for any given Q and R) of the resultant system and denote it with $J_1(x(0))$. Define now $J_2(x(0))$ as the quadratic performance index (with the same parameters Q and R) for the closed-loop system with the controller proposed in this chapter. It is to be noted that to design this controller, the prediction values used in (2.8) are replaced by their corresponding mean values, as explained in Theorem 5. Moreover, define $J_c(x(0))$ as the minimum achievable performance index for the centralized case. According to Theorem 1, $J_2(\bar{x}_0) = J_c(\bar{x}_0)$, which implies that $J_2(\bar{x}_0) < J_1(\bar{x}_0)$. This means that there is a region $\mathscr{R}$ around the point x_0 in the n dimensional space, such that for any $x(0)$ in this region, the inequality $J_2(x(0)) < J_1(x(0))$ holds. On the other hand, if the function $J_2(x(0))$ is smooth around $\bar{x}_0$, the initial state of the system will have a greater likelihood inside the region $\mathscr{R}$ rather than outside of it, in which case the controller* *proposed in this chapter will outperform the one obtained by the method proposed in [16]. It is to be noted that to evaluate the smoothness of the function $J_2(x(0))$ one can use the formula (2.33) to obtain the function $J_2(x(0))$, in a quadratic form, while for the numerical method such as the one in [16], there is no closed-form formula for $J_1(x(0))$ in terms of the initial state. Similar comparison can analogously be made between the method presented in this chapter and the method given in [13], [14] and [15] (first approach discussed in the introduction).*

2.8 Decentralized high-performance cheap control

Consider now the cheap control optimization problem, where it is desired to minimize a quadratic performance index of the following form:

$$J = \int_0^\infty \left(x^T Qx + \varepsilon u^T Ru\right) dt \tag{2.35}$$

where $Q \in \Re^{n\times n}$ and $R \in \Re^{m\times m}$ are positive definite matrices, and ε is a positive number which is chosen sufficiently close to zero for this type of problem. For simplicity and without loss of generality, assume again that Q and R are symmetric. Consider the matrix K_ε such that the feedback law:

$$u(t) = -K_\varepsilon x(t) \tag{2.36}$$

minimizes the performance index (2.35). In the remainder of this section, assume that K given in (2.5) is equal to K_ε. According to Theorem 2, the local controllers given by (2.8) can stabilize the system $\mathscr{S}$, if all modified systems $\mathbf{S}^i$, $i = 2,3,...,\nu$, under the feedback law (2.36) are stable. These conditions are also *almost always* necessary. In sequel, it will be shown that if $det\left(BR^{-1}B^T\right) \neq 0$, and if ε is sufficiently close to zero, there is no need to check the stability of the $\nu-1$ closed-loop modified systems.

Theorem 6 *Assume that $s_1^\varepsilon, s_2^\varepsilon, ..., s_n^\varepsilon$ are the eigenvalues of the system $\mathscr{S}$ under the feedback law $u(t) = -K_\varepsilon x(t)$, and that the determinant of the matrix $BR^{-1}B^T$ is nonzero. Then, as ε approaches zero, $\sqrt{\varepsilon}s_1^\varepsilon, \sqrt{\varepsilon}s_2^\varepsilon, ..., \sqrt{\varepsilon}s_n^\varepsilon$ converge to n negative (nonzero) real numbers $\hat{s}_1, \hat{s}_2, ..., \hat{s}_n$, which satisfy the following equation:*

$$det\left(\hat{s}_i^2 I - WQ\right) = 0, \quad i = 1,2,...,n \tag{2.37}$$

where $W = BR^{-1}B^T$.

Proof It is known that the state and costate of the system $\mathscr{S}$ under the optimal feedback law $u(t) = -K_\varepsilon x(t)$ satisfy the following equation [25]:

$$\begin{bmatrix} \dot{x} \\ \dot{\lambda} \end{bmatrix} = H \times \begin{bmatrix} x \\ \lambda \end{bmatrix}, \quad H = \begin{bmatrix} A & -\frac{W}{\varepsilon} \\ -Q & -A^T \end{bmatrix}, \quad W = BR^{-1}B^T$$

The matrix H has $2n$ eigenvalues in mirror-image pairs with respect to the imaginary axis. Those eigenvalues which are in the left-half s-plane are the eigenvalues of the closed-loop system under the feedback law $u(t) = -K_\varepsilon x(t)$. The eigenvalues of the matrix H are obtained from the following equation:

$$det \begin{bmatrix} sI - A & \frac{W}{\varepsilon} \\ Q & sI + A^T \end{bmatrix} = 0 \tag{2.38}$$

Let the roots of the above equation be denoted by $s_1^\varepsilon, s_2^\varepsilon, ..., s_{2n}^\varepsilon$, where $s_{i+n}^\varepsilon = -s_i^\varepsilon$, $Re\{s_i^\varepsilon\} \leq 0$, for $i = 1, 2, ..., n$. One can multiply the first n rows and the last n columns of the matrix in the left side of (2.38) by $\sqrt{\varepsilon}$ to obtain the following relation:

$$det \begin{bmatrix} sI - A & \frac{W}{\varepsilon} \\ Q & sI + A^T \end{bmatrix} = \frac{1}{\varepsilon^n} det \begin{bmatrix} \sqrt{\varepsilon} sI - \sqrt{\varepsilon} A & W \\ Q & \sqrt{\varepsilon} sI + \sqrt{\varepsilon} A^T \end{bmatrix} \tag{2.39}$$

Hence, it can be concluded from (2.38) and (2.39), that $s_1^\varepsilon, s_2^\varepsilon, ..., s_n^\varepsilon$ are the roots of the following equation:

$$det \begin{bmatrix} \sqrt{\varepsilon} sI - \sqrt{\varepsilon} A & W \\ Q & \sqrt{\varepsilon} sI + \sqrt{\varepsilon} A^T \end{bmatrix} = 0$$

Define $\hat{s}_i^\varepsilon = \sqrt{\varepsilon} s_i^\varepsilon$, $i = 1, 2, ..., n$. Consequently, $\hat{s}_1^\varepsilon, \hat{s}_2^\varepsilon, ..., \hat{s}_n^\varepsilon$ satisfy the following equation:

$$det \begin{bmatrix} sI - \sqrt{\varepsilon} A & W \\ Q & sI + \sqrt{\varepsilon} A^T \end{bmatrix} = 0 \tag{2.40}$$

It can be easily verified that the above equation is equivalent to

$$s^{2n} + p_{2n-1}(\sqrt{\varepsilon}) s^{2n-1} + p_{2n-2}(\sqrt{\varepsilon}) s^{2n-2} + \cdots + p_1(\sqrt{\varepsilon}) s + p_0(\sqrt{\varepsilon}) = 0 \tag{2.41}$$

where $p_i(\sqrt{\varepsilon})$, $i = 1, 2, ..., 2n-1$, is a polynomial in $\sqrt{\varepsilon}$. Obviously, as ε approaches zero, $\hat{s}_1^\varepsilon, \hat{s}_2^\varepsilon, ..., \hat{s}_n^\varepsilon$ (which satisfy the equation (2.40) or equivalently (2.41)), converge to n definite and finite complex numbers denoted by $\hat{s}_1, \hat{s}_2, ..., \hat{s}_n$ (note that the roots of a polynomial equation with finite coefficients are finite), and also they satisfy the equation (2.41) for $\varepsilon = 0$, i.e.

$$\hat{s}_i^{2n} + p_{2n-1}(0)\hat{s}_i^{2n-1} + p_{2n-2}(0)\hat{s}_i^{2n-2} + \cdots + p_1(0)\hat{s}_i + p_0(0) = 0, \quad i = 1, 2, ..., n$$

To find $\hat{s}_i$, replace ε with zero and substitute $s = \hat{s}_i$ in the equation (2.40). This results in:

$$det \begin{bmatrix} \hat{s}_i I & W \\ Q & \hat{s}_i I \end{bmatrix} = 0, \quad i = 1, 2, ..., n$$

The above equation can be simplified as follows:

$$0 = det \begin{bmatrix} \hat{s}_i I & W \\ Q & \hat{s}_i I \end{bmatrix} = \det(\hat{s}_i^2 I - WQ), \quad i = 1, 2, ..., n$$

So far, it has been shown that as ε approaches zero, $\sqrt{\varepsilon} s_1^{\varepsilon}, \sqrt{\varepsilon} s_2^{\varepsilon}, ..., \sqrt{\varepsilon} s_n^{\varepsilon}$ converge to the definite numbers $\hat{s}_1, \hat{s}_2, ..., \hat{s}_n$, which satisfy equation (2.37). Since R is positive definite and symmetric, W is positive definite and symmetric as well (note that $det(W) \neq 0$). Using Cholesky decomposition, one can easily conclude that all of the eigenvalues of the matrix WQ are positive real numbers. Therefore, the equation (2.37) implies that $\hat{s}_1^2, \hat{s}_2^2, ..., \hat{s}_n^2$ are positive real numbers and consequently, $\hat{s}_1, \hat{s}_2, ..., \hat{s}_n$ are purely real. Since the feedback law $u(t) = -K_{\varepsilon} x(t)$ stabilizes the system $\mathscr{S}$, all of the eigenvalues of this closed-loop system are located in the left-half s-plane. As a result, $\hat{s}_1, \hat{s}_2, ..., \hat{s}_n$ are non-positive. Also, it is apparent that none of $\hat{s}_1, \hat{s}_2, ..., \hat{s}_n$ are zero, because in that case $det(WQ) = 0$, which is a contradiction to the assumption of positive definite Q (note that $det(W) \neq 0$). ■

It is to be noted that the result of Theorem 6 is an extension of the existing results for the modes of optimal closed-loop SISO systems and the corresponding inverse root characteristic equation [31], to the MIMO case.

As an example, consider a system consisting of two 2-input 2-output subsystems and the following state-space matrices:

$$A = \begin{bmatrix} 1 & 2 & 0 & 0 \\ -2 & 30 & 0 & 0 \\ 4 & 6 & 1 & 2 \\ -5 & 5 & 7 & 5 \end{bmatrix}, \quad B = \begin{bmatrix} 2 & 60 & 0 & 0 \\ 2 & 6 & 0 & 0 \\ 0 & 0 & 10 & 1 \\ 0 & 0 & 3 & 3 \end{bmatrix} \tag{2.42}$$

Solving the centralized optimal LQR problem for $R = Q = I$ and multiplying the eigenvalues of the resultant closed-loop system (under the feedback law (2.36)) by $\sqrt{\varepsilon}$ as described in Theorem 6, will result in $\{\sqrt{\varepsilon}s_1^{\varepsilon}, \sqrt{\varepsilon}s_2^{\varepsilon}, \sqrt{\varepsilon}s_3^{\varepsilon}, \sqrt{\varepsilon}s_4^{\varepsilon}\}$. The following sets of eigenvalues are obtained for $\varepsilon = 10^{-2}, 10^{-3}, 10^{-4}$ and 10^{-5}, respectively:

$$\begin{aligned} &\{-60.336, -10.609, -3.4573, -2.7356\}, \quad \{-60.339, -10.608, -2.5601, -2.0257\}, \\ &\{-60.339, -10.608, -2.5469, -1.8147\}, \quad \{-60.339, -10.608, -2.5455, -1.7924\} \end{aligned} \tag{2.43}$$

On the other hand, the roots of (2.37) are given by:

$$\{\pm 60.339, \pm 10.608, \pm 2.5453, \pm 1.7899\} \tag{2.44}$$

From (2.43) and (2.44), it is evident that as ε become smaller, the modes of the optimal closed-loop system under the feedback law (2.36) approach the negative elements of the set (2.44), as expected from Theorem 6 (Note that $BR^{-1}B^T$ is nonsingular in this example).

Lemma 1 *Consider two arbitrary symmetric positive-definite matrices G and H. There is a unique positive definite matrix X which satisfies the following relation:*

$$XGX = H$$

Proof It is known that every symmetric positive-definite matrix can be uniquely written as the square of another symmetric positive definite matrix. Therefore, there is a unique positive definite matrix $\hat{G}$ such that $G = \hat{G}^2$. Define $Y = \hat{G}X\hat{G}$, or equivalently $X = \hat{G}^{-1}Y\hat{G}^{-1}$. It is clear that since $\hat{G}$ and X are positive definite and $\hat{G}$ is symmetric, Y is also positive definite, and

$$H = X\hat{G}^2X = \hat{G}^{-1}Y\hat{G}^{-1}\hat{G}^2\hat{G}^{-1}Y\hat{G}^{-1} = \hat{G}^{-1}Y^2\hat{G}^{-1}$$

or equivalently:

$$Y^2 = \hat{G}H\hat{G} \tag{2.45}$$

Similarly, since H and G are positive definite, $\hat{G}H\hat{G}$ is positive definite as well. Therefore, there is a unique positive definite matrix Y whose square is equal to $\hat{G}H\hat{G}$. The matrix Y satisfies the equation (2.45), and thus X is determined to be equal to $\hat{G}^{-1}Y\hat{G}^{-1}$, which is also unique. ■

Theorem 7 *Suppose that the matrix W corresponding to the system (2.1) and the performance index (2.35) is nonsingular. Consider the modified system* $\mathbf{S}^j$, $j \in \{2,3,...,\nu\}$. *There exists a finite* $\varepsilon^* > 0$ *such that for every positive real number* ε *less than* ε^*, *the modified system* $\mathbf{S}^j$ *is stable under the feedback law (2.36).*

Proof Assume that the modes of the system $\mathscr{S}$ under the feedback law (2.36) are $s_1^\varepsilon, s_2^\varepsilon, ..., s_n^\varepsilon$. It is clear that these modes satisfy the following equation:

$$det\left(s_i^\varepsilon I - A + BK_\varepsilon\right) = 0, \quad i = 1,2,...,n \tag{2.46}$$

Suppose that P_ε is the solution of the Riccati equation for the system $\mathscr{S}$ and the performance index (2.35). Thus,

$$-P_\varepsilon A - A^T P_\varepsilon - Q + \frac{1}{\varepsilon} P_\varepsilon BR^{-1}B^T P_\varepsilon = 0 \tag{2.47}$$

Since $K_\varepsilon = \frac{1}{\varepsilon}R^{-1}B^T P_\varepsilon$ and $W = BR^{-1}B^T$, the equation (2.46) can be rewritten as

$$det\left(s_i^\varepsilon I - A + \frac{1}{\varepsilon} W P_\varepsilon\right) = 0, \quad i = 1,2,...,n \tag{2.48}$$

According to Theorem 6, as ε approaches zero, $\sqrt{\varepsilon} s_i^\varepsilon$ converges to the negative definite number $\hat{s}_i$ for $i = 1,2,...,n$. Using this approximation and substituting it into (2.48) will result in (as $\varepsilon \to 0$):

$$det\left(\frac{\hat{s}_i}{\sqrt{\varepsilon}} I - A + \frac{1}{\varepsilon} W P_\varepsilon\right) \to 0, \quad i = 1,2,...,n$$

Define $\hat{P}_\varepsilon := \frac{P_\varepsilon}{\sqrt{\varepsilon}}$. It can then be concluded from the above equation that as ε goes to zero,

$$det\left(\hat{s}_i I - \sqrt{\varepsilon} A + W \hat{P}_\varepsilon\right) \to 0, \quad i = 1,2,...,n \tag{2.49}$$

Substituting $P_\varepsilon = \sqrt{\varepsilon} \hat{P}_\varepsilon$ in the Riccati equation (2.47) yields

$$-\sqrt{\varepsilon} \hat{P}_\varepsilon A - \sqrt{\varepsilon} A^T \hat{P}_\varepsilon - Q + \hat{P}_\varepsilon W \hat{P}_\varepsilon = 0$$

Since the solution of the Riccati equation as well as the matrices W and Q are all positive definite, according to Lemma 1, as ε approaches zero, $\hat{P}_\varepsilon$ converges to a unique positive definite matrix denoted by $\hat{P}$, which can be obtained by solving the equation $Q = \hat{P} W \hat{P}$, as discussed in Lemma 1. In other words, as ε goes to zero, the solution of the Riccati equation P_ε for the system $\mathscr{S}$ and the performance index (2.35) can be estimated by $\sqrt{\varepsilon}\hat{P}$. Accordingly, since $\hat{P}_\varepsilon$ converges to $\hat{P}$ as ε approaches zero, the equation (2.49) yields

$$det\left(\hat{s}_i I + W\hat{P}\right) = 0, \quad i = 1, 2, ..., n \tag{2.50}$$

Now, consider the modified system $\mathbf{S}^j$ under the feedback law (2.36), and let the corresponding closed-loop modes be denoted by $\sigma_{1j}^\varepsilon, \sigma_{2j}^\varepsilon, ..., \sigma_{nj}^\varepsilon$. It is clear that these modes satisfy the following equation:

$$det\left(\sigma_{ij}^\varepsilon I - \tilde{A}^j + \frac{1}{\varepsilon} W P_\varepsilon\right) = 0, \quad i = 1, 2, ..., n \tag{2.51}$$

The above discussion shows that as ε goes to zero, P_ε converges to $\sqrt{\varepsilon}\hat{P}$. Therefore, it can be concluded from the equation (2.51) that (as ε approaches zero):

$$det\left(\sqrt{\varepsilon}\sigma_{ij}^\varepsilon I - \sqrt{\varepsilon}\tilde{A}^j + W\hat{P}\right) \to 0, \quad i = 1, 2, ..., n$$

Since all entries of the matrix $\tilde{A}^j$ are finite and independent of ε, the above expression is equivalent to the following:

$$det\left(\sqrt{\varepsilon}\sigma_{ij}^\varepsilon I + W\hat{P}\right) \to 0, \quad i = 1, 2, ..., n \tag{2.52}$$

By comparing equations (2.50) and (2.52), it can be concluded that as ε goes to zero, the elements of the set $\{\sqrt{\varepsilon}\sigma_{1j}^\varepsilon, ..., \sqrt{\varepsilon}\sigma_{nj}^\varepsilon\}$ converge to the elements of the set $\{\hat{s}_1, ..., \hat{s}_n\}$. According to Theorem 6, $\hat{s}_1, ..., \hat{s}_n$ are all negative numbers. Thus, $\sqrt{\varepsilon}\sigma_{1j}^\varepsilon, ..., \sqrt{\varepsilon}\sigma_{nj}^\varepsilon$ will go to n negative real numbers. As a result, as ε approaches zero, all of these modes will move towards the left-half s-plane, and eventually all of them will be located in the open left-half s-plane. Thus, from continuity, one can conclude that there is a positive value ε^* such that for every ε less than ε^*, all complex numbers $\sigma_{1j}^\varepsilon, ..., \sigma_{nj}^\varepsilon$ will have negative real parts, and hence, the resultant closed-loop system will be stable. ■

Remark 8 *As ε approaches zero, $\sqrt{\varepsilon}\sigma_{1j}^\varepsilon, \sqrt{\varepsilon}\sigma_{2j}^\varepsilon, ..., \sqrt{\varepsilon}\sigma_{nj}^\varepsilon$ converge to n finite negative real numbers. Thus, $\sigma_{1j}^\varepsilon, \sigma_{2j}^\varepsilon, ..., \sigma_{nj}^\varepsilon$ all go to $-\infty$.*

Remark 9 *Since the elements of both sets $\{\sqrt{\varepsilon}\sigma_{1j}^\varepsilon, ..., \sqrt{\varepsilon}\sigma_{nj}^\varepsilon\}$ and $\{\sqrt{\varepsilon}s_1^\varepsilon, ..., \sqrt{\varepsilon}s_n^\varepsilon\}$ approach the elements of the set $\{\hat{s}_1, ..., \hat{s}_n\}$ as ε goes to zero, it can be deduced that the modes of the modified system $\mathbf{S}^j$ under the feedback law (2.36) become closer to the modes of the original system $\mathscr{S}$ under the feedback law (2.36), as ε approaches zero.*

Consider again the system represented by the state-space matrices (2.42). Solving the centralized optimal LQR problem for $R = Q = I$ and multiplying the eigenvalues of the resultant closed-loop modified system $\mathbf{S}^2$ (under the feedback law (2.36)) by $\sqrt{\varepsilon}$ as described in Theorem 7, i.e. $\{\sqrt{\varepsilon}\sigma_{12}^{\varepsilon}, \sqrt{\varepsilon}\sigma_{22}^{\varepsilon}, \sqrt{\varepsilon}\sigma_{32}^{\varepsilon}, \sqrt{\varepsilon}\sigma_{42}^{\varepsilon}\}$, the following results are obtained for $\varepsilon = 10^{-2}, 10^{-3}, 10^{-4}$ and 10^{-5}, respectively:

$$\begin{aligned} &\{-60.467, -10.607, -3.4948, -2.5690\}, \quad \{-60.352, -10.607, -2.0194, -2.5533\}, \\ &\{-60.340, -10.608, -2.5460, -1.8143\}, \quad \{-60.339, -10.608, -2.5454, -1.7923\} \end{aligned} \tag{2.53}$$

Comparing (2.53) and (2.44), it can be seen that as ε goes to zero, the eigenvalues of the modified system $\mathbf{S}^2$ under the feedback law (2.36) multiplied by $\sqrt{\varepsilon}$ converge to the negative elements of the set given by (2.44), as expected.

Corollary 2 *Suppose that $det(W) \neq 0$. If ε is sufficiently close to zero, the system $\mathscr{S}$ under the proposed control law (2.36) is stable.*

Proof It can be concluded from Theorem 7 that there is an ε^* such that for every positive real value $\varepsilon < \varepsilon^*$, all of the systems $\mathbf{S}^2, \mathbf{S}^3, ..., \mathbf{S}^{\nu}$ are stable under the feedback law (2.36). Therefore, according to Theorem 2, the proposed decentralized feedback law stabilizes the system $\mathscr{S}$ (for any $0 < \varepsilon < \varepsilon^*$). ■

Remark 10 *To investigate robust stability of the proposed decentralized cheap control law, one can use the result of Theorem 3 to find the permissible range of parameter variations. As a particular case, assume that $det(W) \neq 0$ and that $\bar{B}_i = B_i$ for $i = 1, 2, ..., \nu$, i.e. there is no perturbation in the entries of the matrix B. It was shown that as ε approaches zero, the modes of the modified system $\mathbf{S}^i$ under the feedback law (2.36) $(\sigma_{1i}^{\varepsilon}, \sigma_{2i}^{\varepsilon}, ..., \sigma_{ni}^{\varepsilon})$ converge to $\frac{1}{\sqrt{\varepsilon}}$ times the numbers $\hat{s}_1, \hat{s}_2, ..., \hat{s}_n$, which are obtained for the given B, R and Q using (2.37). In other words, dependency of the eigenvalues of the modified system $\mathbf{S}^i$ under the feedback law (2.36) on the entries of the matrix A is being reduced, as ε goes to zero. Consider now the modified perturbed system $\bar{\mathbf{S}}^i$. The only difference between $\mathbf{S}^i$ and $\bar{\mathbf{S}}^i$ is in the matrices $\tilde{A}^i$ and $\bar{A}^i$, or more specifically, in A_{ii} and $\bar{A}_{ii}$. Hence, as discussed before, the discrepancy between the modes of $\bar{\mathbf{S}}^i$ and $\mathbf{S}^i$ under the feedback law (2.36) is reduced, as ε approaches zero. This means that as ε goes to zero, the eigenvalues of the perturbed system $\bar{\mathbf{S}}$ under the proposed local controllers become insensitive to the entries of the matrix A.*

Remark 11 *It is to be noted that the condition $det(W) \neq 0$ is equivalent to $det(BB^T) \neq 0$, or equivalently $det(B_iB_i^T) \neq 0$ for any $i \in \bar{\mathcal{V}}$. Therefore, if the number of inputs of any subsystem is less than the number of its outputs, then the matrix $B_iB_i^T$ will be singular, and consequently the condition of Theorem 6 will be violated. Although this condition on the number of inputs of each subsystem can be very restrictive in general, in many practical problems it can be satisfied by adding certain actuators to some of the subsystems, if necessary.*

2.9 Numerical examples

In this section, two examples will be presented. The first one is a numerical example which aims to illustrate some of the procedures developed in the chapter. The second one applies the results obtained in this chapter, to the formation flying problem in [2], and involves simulations.

Example 1 Consider a system $\mathscr{S}_a$ consisting of two SISO subsystems and the following state-space matrices:

$$A = \begin{bmatrix} -1 & 0 \\ -20 & 1 \end{bmatrix} \quad , \quad B = \begin{bmatrix} 1 & 0 \\ 0 & 2 \end{bmatrix}$$

The modified system $\mathbf{S}_a^2$ is obtained by removing the interconnection going to the second subsystem (i.e. by setting the entry -20 of A to zero). Suppose that K_ε is the optimal feedback gain for the system $\mathscr{S}$ and the performance index (2.35), with $R = Q = I$. According to Theorem 7, since $det\left(BR^{-1}B^T\right) \neq 0$, there exists a positive real ε^* such that for every positive $\varepsilon < \varepsilon^*$, the modified system $\mathbf{S}_a^2$ under the feedback law $u(t) = -K_\varepsilon x(t)$ is stable. Computing K_ε for $\varepsilon = 1$, the eigenvalues of the modified system $\mathbf{S}_a^2$ under the feedback law $u(t) = -K_\varepsilon x(t)$ are obtained to be 0.2169 and -7.1087. According to Theorem 2, since one of these eigenvalues is positive, the overall closed-loop system is unstable. Therefore, ε^* has to be less than one. It can be verified that for this example $\varepsilon^* \simeq 0.668$. Hence, for every $\varepsilon < 0.668$, the proposed local controllers can stabilize the system $\mathscr{S}_a$.

Let ε be equal to 0.001 . Computing K_ε for this value of ε, it can be shown that the eigenvalues of the system $\mathscr{S}_a$ and the modified system $\mathbf{S}_a^2$ under the feedback law $u(t) = -K_\varepsilon x(t)$ are $\{-62.1381, -33.7765\}$ and $\{-64.3194, -31.5952\}$, respectively. The eigenvalues of these two closed-loop systems are close to each other as pointed out in Remark 9.

In the next step, it is desired to inspect the robustness of the system $\mathscr{S}_a$ under the proposed decentralized control law for $\varepsilon = 0.001$, and compare it to the robustness of the system $\mathscr{S}_a$ under the centralized feedback law $u(t) = -K_\varepsilon x(t)$.

1. *Decentralized case:* According to Theorem 3, the perturbed system $\bar{\mathscr{S}}_a$ under the proposed decentralized controller is stable if the modified perturbed systems $\bar{\mathbf{S}}_a^1$ and $\bar{\mathbf{S}}_a^2$ under the feedback law $u(t) = -K_\varepsilon x(t)$ are both stable. Therefore, any s which satisfies one of the following equations:

$$\det\left(sI - \bar{A}^1 + \bar{B}^1 K\right) = 0, \quad \det\left(sI - \bar{A}^2 + \bar{B}^2 K\right) = 0 \tag{2.54}$$

should have a negative real part. It is desired now to find some relations which exhibit the maximum allowable deviations from the nominal parameters of the system. Define:

$$\Delta A_{ij} := \bar{A}_{ij} - A_{ij}, \quad i,j \in \{1,2\}, \quad i \geq j$$

$$\Delta B_i := \bar{B}_i - B_i, \quad i = 1,2$$

Since all of the roots of the equations given in (2.54) should be in the left-half s-plane, it is easy to verify that the allowable perturbations are given by the following inequalities:

$$\begin{gathered} 32.366\Delta B_1 - \Delta A_{11} > -33.366, \quad 32.013\Delta B_1 - \Delta A_{11} > -95.915, \\ 31.951\Delta B_2 - \Delta A_{22} > -95.915, \quad -31.278\Delta B_2 - \Delta A_{22} > -61.557, \\ \Delta A_{21} = arbitrary \end{gathered} \tag{2.55}$$

2. *Centralized case:* Consider the perturbed system $\bar{\mathscr{S}}_a$ under the feedback law $u(t) = -K_\varepsilon x(t)$. The closed-loop system is stable, iff all of the roots of the equation $det\left(sI - \bar{A} + \bar{B}K\right) = 0$ have negative real parts. Hence, the allowable ranges of perturbations in the centralized case satisfy the following inequalities:

$$-9.910\Delta A_{22} - 18.88\Delta A_{11} - \Delta A_{21} + 611.163\Delta B_1 + 309.997\Delta B_2 + 0.3\Delta A_{11}\Delta A_{22}$$
$$-9.591\Delta A_{11}\Delta B_2 - 9.610\Delta A_{22}\Delta B_1 - \Delta A_{21}\Delta B_1 + 300.386\Delta B_1\Delta B_2 > -630.045\,,$$
$$-\Delta A_{11} - \Delta A_{22} + 32.013\Delta B_1 + 31.951\Delta B_2 > -95.915$$

To compare robustness of the decentralized and centralized controllers, suppose that $\Delta B_1 = \Delta B_2 = 0$. According to the inequalities in (2.55), the admissible parameter variations in the decentralized case are as follows:

$$\Delta A_{11} < 33.366\,,\ \Delta A_{22} < 61.557\,,\ \Delta A_{21} < +\infty \tag{2.56}$$

The admissible parameter variations in the centralized case, on the other hand, are given by:

$$9.910\Delta A_{22} + 18.88\Delta A_{11} + \Delta A_{21} - 0.3\Delta A_{11}\Delta A_{22} < 630.045 \tag{2.57a}$$

$$\Delta A_{11} + \Delta A_{22} < 95.915 \tag{2.57b}$$

From (2.56) and (2.57), it is clear that the centralized controller is less robust to the parameter variations compared to its decentralized counterpart, because:

- Stability in the decentralized case is independent of ΔA_{21} but in the centralized case it is not.
- Regardless of ΔA_{21}, there is the term $\Delta A_{11}\Delta A_{22}$ in the centralized case. This implies that when the two perturbations ΔA_{11} and ΔA_{22} have the same sign, (2.57a) can be easily violated, even if ΔA_{21} is zero.

Example 2 Consider a leader-follower formation control system consisting of three unmanned aerial vehicles. It is known that each vehicle, except the leader, can measure its relative position with respect to the preceding vehicle by using a GPS based architecture [2]. Therefore, it is assumed in this example that each vehicle is equipped with this measuring device, and that the velocity and the acceleration of each vehicle are not available for the other vehicles due to the information exchange constraint between the vehicles. The objective is to design three local controllers for these vehicles, such that they all fly at the same desired speed, with the prespecified desired Euclidean distances between them. It is shown in [2] that using the exact linearization technique, the tracking system with the normalized parameters can be modeled as a regulation problem with the following state-space representation:

$$\begin{bmatrix} \dot{x}_1 \\ \dot{x}_2 \\ \dot{x}_3 \end{bmatrix} = \begin{bmatrix} 0_2 & 0_2 & 0_2 & 0_2 & 0_2 \\ I_2 & 0_2 & -I_2 & 0_2 & 0_2 \\ 0_2 & 0_2 & 0_2 & 0_2 & 0_2 \\ 0_2 & 0_2 & I_2 & 0_2 & -I_2 \\ 0_2 & 0_2 & 0_2 & 0_2 & 0_2 \end{bmatrix} \begin{bmatrix} x_1 \\ x_2 \\ x_3 \end{bmatrix} + \begin{bmatrix} I_2 & 0_2 & 0_2 \\ 0_2 & 0_2 & 0_2 \\ 0_2 & I_2 & 0_2 \\ 0_2 & 0_2 & 0_2 \\ 0_2 & 0_2 & I_2 \end{bmatrix} \begin{bmatrix} u_1 \\ u_2 \\ u_3 \end{bmatrix} \tag{2.58}$$

where I_2 and 0_2 represent a 2×2 identity matrix and a 2×2 zero matrix, respectively, and

$$x_1 = \begin{bmatrix} x_{11} \\ x_{12} \end{bmatrix}, \; x_2 = \begin{bmatrix} x_{21} \\ x_{22} \\ x_{23} \\ x_{24} \end{bmatrix}, \; x_3 = \begin{bmatrix} x_{31} \\ x_{32} \\ x_{33} \\ x_{34} \end{bmatrix} \tag{2.59}$$

and where $u_i = \begin{bmatrix} u_{i1} \\ u_{i2} \end{bmatrix}$, $i = 1,2,3$. Here, x_1 denotes the state of the leader, and x_2 and x_3 represent the state of vehicles 2 and 3 (i.e., the followers), respectively. More specifically:

1. x_{11} and x_{12} are the speed error of the leader (speed of the leader minus its desired speed) along x and y axes, respectively.

2. x_{i1} and x_{i2}, $i = 2,3$, are the distance error (distance between vehicles i and $i-1$ minus their desired distance) along x and y axes, respectively.

3. x_{i3} and x_{i4}, $i = 2,3$, are the speed error (speed of vehicle i minus its desired speed) along x and y axes, respectively.

4. u_{i1} and u_{i2}, $i = 1,2,3$, are the acceleration of vehicle i along x and y axes, respectively.

It is desired now to design a decentralized controller for the system given by (2.58), such that the closed-loop system is stable. Moreover, the objective is that the state variables of the closed-loop system decay as sharply as possible, with a reasonably small control effort. To attain these specifications, consider the performance index given by (2.3) in the chapter, and assume that $Q = R = I$. Two different design techniques will be used and the results will be compared here: the iterative numerical procedure given in [16], and the method proposed in this chapter. Suppose that each initial state is uniformly distributed in the intervals $[200, 400]$, and that any two distinct initial state variables are statistically independent. It is to be noted that the units used for distance and velocity in the state vectors are ft and ft/s, respectively. Assume that any two different subsystems consider the same expected value for the initial state of the remaining subsystem, and that the model of each subsystem is exactly known by the other subsystems. It can be concluded from Procedure 1 and Remark 7 that if the real initial state variables are close to their expected value 300, the controller obtained by using the proposed method performs better.

Assume that the real initial state variables are all equal to 400, which correspond, in fact, to the worst case scenario (maximum discrepancy between the real initial state variables, i.e. 400, and the corresponding expected values, i.e. 300, which are used by the proposed controller). The iterative numerical procedure of [16] gives a static decentralized state feedback law which results in a performance index equal to 2,257,085. The performance index obtained by applying the method proposed in this chapter, on the other hand, is equal to 2,090,939, while the best achievable performance index corresponding to the centralized LQR controller is equal to 2,068,513. This means that the relative errors of the performance indices obtained by using the methods given here and in [16], with respect to the optimal centralized performance index are 1.08% and 9.12%, respectively. This shows clearly that the controller proposed in this chapter outperforms the one presented in [16], significantly.

Figures 2.1 and 2.2 depict the time responses of the system under the controller proposed in this chapter (dotted curve), the controller proposed in [16] (dashed curve), and the optimal centralized controller (solid curve) for three state variables x_{11}, x_{31}, x_{33}. Moreover, the control signals u_{11}, u_{21}, u_{31} obtained by using the three methods discussed above are depicted in Figures 2.2 and 2.3 in a similar way. It is to be noted that despite the relatively big differences between the real initial variables (400 ft for distance errors and 400 ft/sec for speed errors) and the corresponding expected values which are used to construct the proposed controller, the results obtained are reasonably close to the time response of the system under the LQR controller.

The results obtained show that the method introduced in present work is much better than the one in [16]. On the other hand, as stated in the introduction, the controller obtained by the method in [16] has a better performance compared to the ones proposed in [13], [14], and [15]. In addition, the control law given in [23] can potentially outperform the ones in [13], [14], and [15], but can never perform better than the one in [16]. This exhibits superiority of the proposed design technique over the existing ones.

2.10 Conclusions

In this chapter, an incrementally linear decentralized control law for the formation of vehicles with leader-follower structure is introduced. The fundamental idea in constructing this control law is that the local controller of each vehicle exploits *a priori* information about the models and the expected

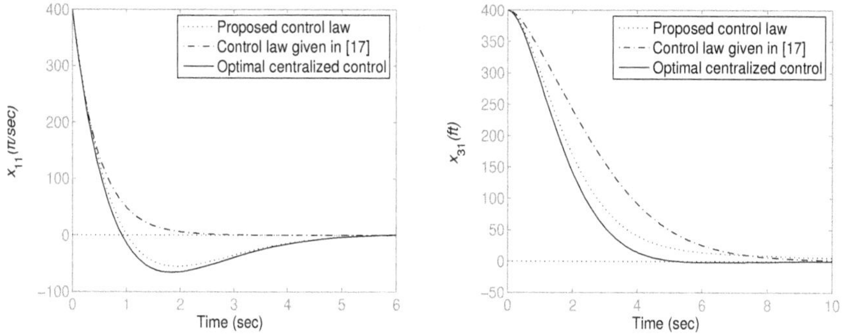

Figure 2.1: The state variables x_{11} and x_{31} using three different design techniques.

values of all other vehicles. It is shown that the decentralized closed-loop system can behave the same as the optimal centralized closed-loop system (with respect to a quadratic performance index) if the *a priori* knowledge of each subsystem is perfect. Since this knowledge can be inaccurate, the performance degradation of the resultant decentralized closed-loop system has been evaluated thoroughly, in presence of inexact information. The proposed decentralized control strategy is very easy to implement, and the corresponding stability verification steps are very easy to check as illustrated in the examples. Furthermore, it is shown that the decentralized control system is, in general, more robust than its centralized counterpart. Optimal decentralized cheap control problem is investigated for leader-follower formation structure, and a closed-form solution is provided for the case when the input structure meets a certain condition. This can be very useful for *UAV* missions with fast tracking objectives. Simulation results demonstrate the effectiveness of the proposed method compared to the existing ones.

Bibliography

[1] H. G. Tanner, G. J. Pappas and V. Kumar, "Leader to formation stability," *IEEE Trans. Robot. Automat.*, vol. 20, no. 3, pp. 443-455, Jun. 2004.

[2] D. M. Stipanovi, G. Inalhan, R. Teo and C. J. Tomlin, "Decentralized overlapping control of a

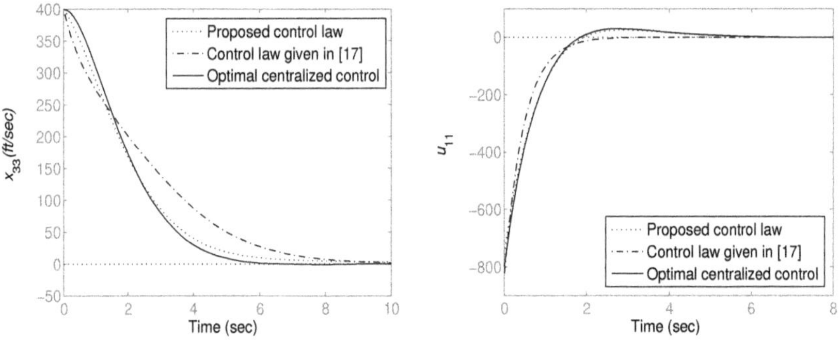

Figure 2.2: The state variable x_{33} and the control signal u_{11} using three different design techniques.

formation of unmanned aerial vehicles," *Automatica*, vol. 40 , no.8, pp. 1285-1296, Aug. 2004.

[3] S. S. Stankovic, M. J. Stanojevic and D. D. Šiljak, "Decentralized overlapping control of a platoon of vehicles," *IEEE Trans. Contrl. Sys. Tech.*, vol. 8, no. 5, pp. 816-832, Sep. 2000.

[4] H. G. Tanner, G. J. Pappas and V. Kumar, "Input-to-state stability on formation graphs," *in Proc. 41st IEEE Conf. on Decision and Contrl.*, Las Vegas, NV, vol. 3, pp. 2439-2444, 2002.

[5] R. L. Raffard, C. J. Tomlin and S. P. Boyd, "Distributed optimization for cooperative agents: application to formation flight," *in Proc. 43rd IEEE Conf. on Decision and Contrl.*, Atlantis, Bahamas, vol. 3, pp. 2453-2459, 2004.

[6] J. A. Fax and R. M. Murray, "Information flow and cooperative control of vehicle formations," *IEEE Trans. Automat. Contrl.*, vol. 49, no. 9, pp. 1465-1476, Sep. 2004.

[7] D. J. Stilwell and B. E. Bishop, "Platoons of underwater vehicles," *IEEE Contrl. Sys. Mag.*, vol. 20, no. 6, pp. 45-52, Dec. 2000.

[8] B. J. Naasz, C. D. Karlgaard and C. D. Hall, "Application of several control techniques for the ionospheric observation nanosatellite," *2002 in Proc. AAS/AIAA Space Flight Mechanics Conf.*, San Antonio, TX, Jan 27-30, 2002.

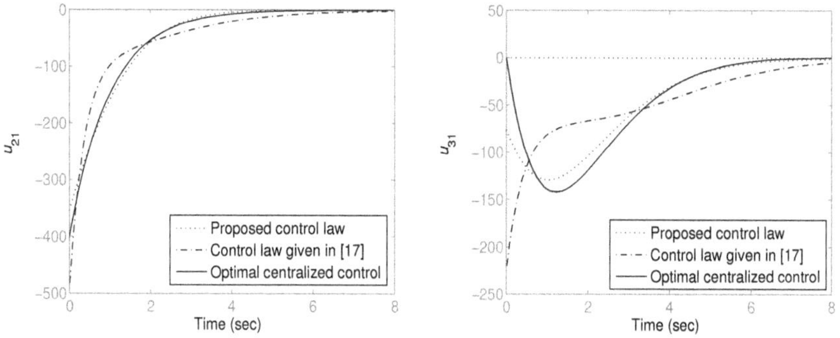

Figure 2.3: The control signals u_{21} and u_{31} using three different design techniques.

[9] G. Inalhan, D. M. Stipanovic and C. J. Tomlin, "Decentralized optimization with application to multiple aircraft coordination," *in Proc. 41st IEEE Conf. on Decision and Contrl.*, Las Vegas, NV, vol. 1, pp. 1147-1155, 2002.

[10] F. Giulietti, L. Pollini and M. Innocenti, "Autonomous formation flight," *IEEE Contrl. Sys. Mag.*, vol. 20, no. 6, pp. 34-44, Dec. 2000.

[11] K. Lee, "An optimal control for multiple unmanned underwater crawling vehicles," *in Proc. of Unmanned Ultethered Submersible Tech.*, 2003.

[12] W. Ren and R. W. Beard, "Formation feedback control for multiple spacecraft via virtual structures," *in Proc. 2004 IEEE of Control Theory Appl.*, vol. 151, no. 3, pp. 357-368, 2004.

[13] R. Krtolica and D. D. Šiljak, "Suboptimality of decentralized stochastic control and estimation," *IEEE Trans. Automat. Contrl.*, vol. 25, no. 1, pp. 76-83, Feb. 1980.

[14] K. D. Young, "On near optimal decentralized control," *Automatica*, vol. 21, no. 5, pp. 607-610, Sep. 1985.

[15] D. D. Šiljak, Decentralized control of complex systems, Boston: Academic Press, 1991.

[16] H. T. Toivoneh and P. M. Makila, "A descent anderson-moore algorithm for optimal decentralized control," *Automatica*, vol. 21, no. 6, pp. 743-744, Nov. 1985.

[17] J. R. Broussard, "An approach to the optimal output feedback initial stabilizing gain problem," *in Proc. 29th IEEE Conf. on Decision and Control*, Honolulu, HI, vol. 5, pp. 2918-2920, 1990.

[18] W. S. Levine and M. Athans, "On the determination of the optimal constant output feedback gains for linear multivariable systems," *IEEE Trans. Automat. Contrl.*, vol. 15, no. 1, pp. 44-48, Feb. 1970.

[19] P. M. Makila and H. T. Toivoneh, "Computational methods for parametric LQ problems- p survey," *IEEE Trans. Automat. Contrl.*, vol. 32, no. 8, pp. 658-671, Aug. 1987.

[20] S. V. Savastuk and D. D. Šiljak, "Optimal decentralized control," *in Proc. 1994 American Contrl. Conf.*, Baltimore, MD, vol. 3, pp. 3369-3373, 1994.

[21] D. D. Sourlas and V. Manousiouthakis, "Best achievable decentralized performance," *IEEE Trans. Automat. Contrl.*, vol. 40, no. 11, pp. 1858-1871, Nov. 1995.

[22] Ü. Özgüner and W. R. Perkins, "Optimal control of multilevel large-scale systems," *Int. Jour. Contrl.*, vol. 28, no. 6, pp. 967-980, 1978.

[23] S. S. Stankovic and D. D. Šiljak, "Sequential LQG optimization of hierarchically structured systems," *Automatica*, vol. 25 , no. 4, pp. 545-559, Jul. 1989.

[24] R. S. Smith and F. Y. Hadaegh, "Control Topologies for Deep Space Formation Flying Spacecraft," *in Proc. 2002 American Contrl. Conf.*, Anchorage, AK, vol. 3, pp. 2836-2841, 2002.

[25] D. E. Kirk, Optimal Control Theory, Dover Publication, Mineola, New york, 2004.

[26] E. J. Davison and S. H. Wang, "Properties of linear time-invariant multivariable systems subject to arbitrary output and state feedback," *IEEE Trans. Automat. Contrl.*, vol. 18 , no.1, pp. 24-32, Feb. 1973.

[27] F. M. Callier, W. S. Chan and C. A. Desoer, "Input-output stability of interconnected systems using decompositions: an improved formulation," *IEEE Trans. Automat. Contrl.*, vol. 23 , no.2, pp. 150-163, Apr. 1978.

[28] M. Embree and L. N. Trefethen, "Generalizing eigenvalue theorems to pseudospectra theorems," *SIAM I. Sci. Comput.*, vol. 23, no. 2, pp. 583-590, 2001.

[29] S. C. Eisenstat and I. C. F. Ipsen, "Three absolute perturbation bounds for matrix eigenvalues imply relative bounds," *SIAM J. Matrix Anal. Appl.*, vol. 20, no. 1, pp. 149-158, 1998.

[30] Jos. L. M. Van Dorsselaer, "Several concepts to investigate strongly nonnormal eigenvalue problems," *SIAM I. Sci. Comput.*, vol. 24, no. 3, pp. 1031-1053, 2003.

[31] B. D. O. Anderson and J. B. Moore, Optimal control, Prentice-Hall, N.J., 1989.

Chapter 3

High-Performance Decentralized Control Law Based on a Centralized Reference Controller

3.1 Abstract

This work deals with the decentralized control of interconnected systems, where each subsystem has models of all other subsystems (subject to uncertainty). A decentralized controller is constructed based on a reference centralized controller. It is shown that when *a priori* knowledge of each subsystem about the other subsystems' models is exact, then the decentralized closed-loop system can perform exactly the same as its centralized counterpart. An easy-to-check necessary and sufficient condition for the internal stability of the decentralized closed-loop system is obtained. Moreover, the stability of the closed-loop system in presence of the perturbation in the parameters of the system is investigated, and it is shown that the decentralized control system is likely more robust than its centralized counterpart. A proper cost function is then defined to measure the closeness of the decentralized closed-loop system to its centralized counterpart. This enables the designer to obtain the maximum allowable standard deviation for the modeling errors of the subsystems to achieve a satisfactorily small performance deviation with a sufficiently high probability. The effectiveness of the proposed method is demonstrated in one numerical example.

3.2 Introduction

In control of large-scale interconnected systems, it is often desired to have some form of decentralization. Typical interconnected control systems have several local control stations, which observe only local outputs and control only local inputs, according to the prescribed restrictions in the information flow structure. All the controllers are involved, however, in controlling the overall system.

In the past several years, the problem of decentralized control design has been investigated extensively in the literature [1-9]. These works have studied decentralized control problem from two different viewpoints as follows:

1. The local input and output information of any subsystem is private, and is not accessible by other subsystems [3; 5]. In this case, each local controller should attempt to attenuate the degrading effect of the incoming interconnections on its corresponding subsystem, in addition to contributing to the performance of the overall control system.

2. All output measurements cannot be transmitted to every local control station. Problems of this kind appear, for example, in electric power systems, communication networks, flight formation, robotic systems, to name only a few. In this case, each subsystem can have certain beliefs about other subsystems' models. A local controller is then designed for each subsystem, based on this *a priori* knowledge.

More recently, the problem of optimal decentralized control design has been studied to obtain a high-performance control law with some prescribed constraints for the system. The main objective in this problem is to find a decentralized feedback law for an interconnected system in order to attain a sufficiently small performance index. This problem has been investigated in the literature from the two different viewpoints discussed above [3; 6; 7; 8; 9]. However, there is no efficient approach currently to address the problem from the second viewpoint. The relevant works often attempt to present a near-optimal decentralized controller instead of an optimal one. Furthermore, it is often assumed that the decentralized controller to be designed is static [7]. This assumption can significantly degrade the performance of the overall system. In other words, the overall performance of the system can be improved considerably, if a dynamic feedback law is used instead of a static one.

This chapter deals with the decentralized control problem from the second viewpoint discussed above. Hence, it is assumed that each subsystem of the interconnected system has some beliefs about the parameters of the other subsystems as well as their initial states. A decentralized controller is then constructed based on a given reference centralized controller which satisfies the design specifications. This decentralized control law relies on the expected values of any subsystem's initial state from any other subsystem's view and the beliefs of each subsystem about the other subsystems' parameters.

Some important issues regarding the proposed decentralized control law are also studied in this work. First, an easy-to-check necessary and sufficient condition for the internal stability of the interconnected system under the proposed decentralized control law is presented. Note that although the expected values of the initial sates contribute to the structures of the local controllers, it is shown that they never affect the internal stability of the overall system. Second, it is shown that if the knowledge of any subsystem about the other subsystems' parameters is accurate, then the decentralized closed-loop system can perform exactly the same as the centralized closed-loop system. However, since the exact knowledge of the subsystem's parameters is very unlikely to be available in practice, a performance index is defined to evaluate the closeness of the proposed decentralized closed-loop system to its centralized counterpart. This performance index enables the designer to statistically analyze the closeness of the decentralized control system to its corresponding centralized counterpart. In addition, the stability of the decentralized closed-loop system in presence of perturbation in the parameters of the system is studied. Finally, a near-optimal decentralized control law is introduced, and its properties are investigated. The effectiveness of the proposed method is demonstrated in a numerical example, where it is shown that even in presence of large error in the beliefs of the subsystems, the performance of the proposed decentralized control law is very close to that of the reference centralized controller.

3.3 Model-based decentralized control

Consider a stabilizable LTI interconnected system $\mathscr{S}$ consisting of ν subsystems $S_1, S_2, \ldots, S_\nu$. Suppose that the state-space equation for the system $\mathscr{S}$ is as follows:

$$\begin{aligned} \dot{x}(t) &= Ax(t) + Bu(t) \\ y(t) &= Cx(t) \end{aligned} \tag{3.1}$$

where $x \in \Re^n$, $u \in \Re^m$, $y \in \Re^r$ are the state, the input, and the output of the system $\mathscr{S}$, respectively. Let

$$u(t) := \begin{bmatrix} u_1(t)^T & u_2(t)^T & \dots & u_\nu(t)^T \end{bmatrix}^T, \tag{3.2a}$$

$$x(t) := \begin{bmatrix} x_1(t)^T & x_2(t)^T & \dots & x_\nu(t)^T \end{bmatrix}^T, \tag{3.2b}$$

$$y(t) := \begin{bmatrix} y_1(t)^T & y_2(t)^T & \dots & y_\nu(t)^T \end{bmatrix}^T \tag{3.2c}$$

where $x_i \in \Re^{n_i}$, $u_i \in \Re^{m_i}$, $y_i \in \Re^{r_i}$, $i \in \bar{\nu} := \{1, 2, ..., \nu\}$, are the state, the input, and the output of the ith subsystem S_i, respectively. Let for any $i, j \in \bar{\nu}$, the (i, j) block entry of the matrix A be denoted by $A_{ij} \in \Re^{n_i \times n_j}$. Assume that the matrices B and C are block diagonal with the (i, i) block entries $B_i \in \Re^{n_i \times m_i}$ and $C_i \in \Re^{r_i \times n_i}$, respectively, for any $i \in \bar{\nu}$. Assume also that a centralized LTI controller K_c is designed for the system $\mathscr{S}$, which satisfies the desired design specifications, and denote its state-space equation as follows:

$$\begin{aligned} \dot{z}(t) &= A_k z(t) + B_k y(t) \\ u(t) &= C_k z(t) + D_k y(t) \end{aligned} \tag{3.3}$$

where $z \in \Re^\eta$ is the state of the controller. It is desirable now to design a decentralized controller for the system $\mathscr{S}$ so that it performs approximately (and under certain conditions, exactly) the same as the centralized controller K_c. The following definitions are used to obtain the desired decentralized control law.

Definition 1 *Define $x_0^{i,j}$, $i, j \in \bar{\nu}$, $i \neq j$, as the expected value of the initial state of the subsystem S_i from the viewpoint of the subsystem S_j.*

Definition 2 *For any $i, j \in \bar{\nu}$, $i \neq j$, the modified subsystem S_i^j is defined to be a system obtained from the subsystem S_i by replacing the parameters A_{il}, B_i, C_i of its state equations with A_{il}^j, B_i^j, C_i^j, respectively, for any $l \in \bar{\nu}$. In addition, the initial state of S_i^j is $x_0^{i,j}$ (instead of $x_i(0)$). Denote the state, the input, and the output of the modified subsystem S_i^j with x_i^j, u_i^j and y_i^j, respectively.*

Note that S_i^j denotes the belief of subsystem j about the model of subsystem i. In other words, ideally, it is desired to have $S_i^j = S_i$, for all $j \in \bar{\nu}$, $j \neq i$.

Definition 3 *For any $i \in \bar{\nu}$, $\mathscr{S}^i$ is defined to be a system obtained from $\mathscr{S}$, after replacing each subsystem S_j with the corresponding modified subsystem S_j^i, $j = 1, 2, ..., i-1, i+1, ..., \nu$.*

It is be noted that $\mathscr{S}^i$ represents the belief of subsystem i about the model of the system $\mathscr{S}$.

Definition 4 *Consider the system $\mathscr{S}^i$ under the centralized feedback law (3.3). Remove all interconnections going to the ith subsystem S_i from the other subsystems to obtain a new closed-loop system. One can decompose this closed-loop system to two portions:*

i) the system S_i;

ii) all other interconnected components (consisting of S^i_j, $j = 1,2,...,i-1,i+1,...,\nu$, and K_c, and the corresponding interconnections). Define this interconnected system as augmented local controller for the subsystem S_i, and denote it with K_{d_i}. Denote the state of the controller K_c (inside the controller K_{d_i}) with z^i.

Define the controller consisting of all augmented local controllers $K_{d_1}, K_{d_2}, ..., K_{d_\nu}$ as the augmented decentralized controller K_d. Let this decentralized controller be applied to the interconnected system $\mathscr{S}$ (K_{d_i} to S_i, for all $i \in \bar{\nu}$). Note that to obtain the augmented local controller K_{d_i}, in addition to the output y_i, all of the interconnections coming out of the subsystem S_i are also assumed to be available for the local controller K_{d_i}. However, since these interconnections contain, in fact, the information of the subsystem S_i, this is a feasible assumption in many practical problems, e.g. flight formation and power systems, and does not contradict the requirement of the decentralized information flow for the control law. The following theorem presents one of the main properties of the proposed decentralized controller.

Theorem 1 *Consider the interconnected system $\mathscr{S}$ with the augmented decentralized controller K_d. If $A^j_{il} = A_{il}$, $B^j_i = B_i$, $C^j_i = C_i$, and $x_0^{i,j} = x_i(0)$ for all $i, j, l \in \bar{\nu}$, $i \neq j$, then the input, the output, and the state of the overall decentralized closed-loop system will be the same as those of the system $\mathscr{S}$ under the centralized controller K_c.*

Proof Consider system $\mathscr{S}$ under the decentralized controller K_d. One can easily write the following equations for the state of the subsystem S_i under its augmented local controller K_{d_i}:

$$\begin{aligned}
\dot{x}^i(t) &= \tilde{A}^i x^i(t) + B^i u^i(t) + \sum_{j \in \bar{\nu}, j \neq i} \tilde{A}^{ij} x^j(t) \\
y^i(t) &= C^i x^i(t) \\
\dot{z}^i(t) &= A_k z^i(t) + B_k y^i(t) \\
u^i(t) &= C_k z^i(t) + D_k y^i(t)
\end{aligned} \tag{3.4}$$

where

$$x^i := \begin{bmatrix} x_1^{i\,T} & \dots & x_{i-1}^{i\;\,T} & x_i & x_{i+1}^{i\;\,T} & \cdots & x_\nu^{i\,T} \end{bmatrix}^T$$
$$y^i := \begin{bmatrix} y_1^{i\,T} & \dots & y_{i-1}^{i\;\,T} & y_i & y_{i+1}^{i\;\,T} & \cdots & y_\nu^{i\,T} \end{bmatrix}^T \tag{3.5}$$
$$u^i := \begin{bmatrix} u_1^{i\,T} & \dots & u_{i-1}^{i\;\,T} & u_i & u_{i+1}^{i\;\,T} & \cdots & u_\nu^{i\,T} \end{bmatrix}^T$$

and where

1. $\tilde{A}^i$ is obtained from A by replacing all of the entries of its ith block row except A_{ii} (i.e. $A_{ij}, j \in \bar{\nu}, i \neq j$) with zeros, and replacing its entry A_{lj} with A^i_{lj} for any $l, j \in \bar{\nu},\ l \neq i$.

2. $\tilde{A}^{ij}$ $(i \neq j)$ is obtained from A by setting all of its block entries except A_{ij}, to zero.

3. B^i and C^i are obtained from B and C, after replacing their entries B_l and C_l with B^i_l and C^i_l, respectively, for $l = 1, 2, ..., i-1, i+1, ..., \nu$.

The equation (3.4) can be rewritten in a matrix form as follows:

$$\begin{bmatrix} \dot{x}^i(t) \\ \dot{z}^i(t) \end{bmatrix} = \begin{bmatrix} \tilde{A}^i + B^i D_k C^i & B^i C_k \\ B_k C^i & A_k \end{bmatrix} \begin{bmatrix} x^i(t) \\ z^i(t) \end{bmatrix} + \sum_{j \in \bar{\nu}, j \neq i} \begin{bmatrix} \tilde{A}^{ij} & 0_{n \times \eta} \\ 0_{\eta \times n} & 0_{\eta \times \eta} \end{bmatrix} \begin{bmatrix} x^j(t) \\ z^j(t) \end{bmatrix} \tag{3.6}$$

So far, the states of the subsystem S_i and its corresponding augmented local controller K_{d_i} are obtained. In order to simplify (3.6), the following vector and matrices are defined:

$$x^i_g(t) := \begin{bmatrix} x^i(t) \\ z^i(t) \end{bmatrix}, \quad \tilde{A}^i_g := \begin{bmatrix} \tilde{A}^i + B^i D_k C^i & B^i C_k \\ B_k C^i & A_k \end{bmatrix}, \quad \tilde{A}^{ij}_g := \begin{bmatrix} \tilde{A}^{ij} & 0_{n \times \eta} \\ 0_{\eta \times n} & 0_{\eta \times \eta} \end{bmatrix} \tag{3.7}$$

for any $i, j \in \bar{\nu},\ i \neq j$. Note that the subscript "g" denotes the parameters and variables of the augmented equations. According to (3.6) and (3.7), the state of the system $\mathscr{S}$ under the proposed decentralized controller K_d satisfies the following equation:

$$\dot{x}_g(t) = A^d_g x_g(t) \tag{3.8}$$

where

$$x_g(t) := \begin{bmatrix} x^1_g(t) \\ x^2_g(t) \\ \vdots \\ x^\nu_g(t) \end{bmatrix}, \quad A^d_g := \begin{bmatrix} \tilde{A}^1_g & \tilde{A}^{12}_g & \cdots & \tilde{A}^{1\nu}_g \\ \tilde{A}^{21}_g & \tilde{A}^2_g & \cdots & \tilde{A}^{2\nu}_g \\ \vdots & \vdots & \ddots & \vdots \\ \tilde{A}^{\nu 1}_g & \tilde{A}^{\nu 2}_g & \cdots & \tilde{A}^\nu_g \end{bmatrix} \tag{3.9}$$

and where the subscript "d" represents the decentralized nature of the control system. On the other hand, one can easily verify that the state of the system $\mathscr{S}$ under the centralized controller K_c satisfies the following:

$$\dot{x}_g(t) = A_g^c x_g(t) \tag{3.10}$$

where

$$A_g^c := \begin{bmatrix} A + BD_kC & BC_k \\ B_kC & A_k \end{bmatrix} \tag{3.11}$$

Since it is assumed that $A_{il}^j = A_{il}$, $B_i^j = B_i$, and $C_i^j = C_i$ for all $i, j, l \in \bar{\nu}$, $i \neq j$, one can write:

$$\tilde{A}_g^i = A_g^c - \sum_{j=1, j\neq i}^{\nu} \tilde{A}_g^{ij} \tag{3.12}$$

According to (3.9) and (3.12), A_g^d can be written as the summation of two components Θ and Γ, where

$$\Theta := \begin{bmatrix} A_g^c & 0 & \cdots & 0 \\ 0 & A_g^c & \cdots & 0 \\ \vdots & \vdots & \ddots & \vdots \\ 0 & 0 & \cdots & A_g^c \end{bmatrix},$$
$$\Gamma := \begin{bmatrix} -\sum_{j=1, j\neq 1}^{\nu} \tilde{A}_g^{1j} & \tilde{A}_g^{12} & \cdots & \tilde{A}_g^{1\nu} \\ \tilde{A}_g^{21} & -\sum_{j=1, j\neq 2}^{\nu} \tilde{A}_g^{2j} & \cdots & \tilde{A}_g^{2\nu} \\ \vdots & \vdots & \ddots & \vdots \\ \tilde{A}_g^{\nu 1} & \tilde{A}_g^{\nu 2} & \cdots & -\sum_{j=1, j\neq \nu}^{\nu} \tilde{A}_g^{\nu j} \end{bmatrix} \tag{3.13}$$

Note that the matrix Θ consists of ν block matrices A_g^c on its main diagonal and that each of the diagonal block entries of Γ is equal to minus the summation of the other block entries of its own block row. Consider now the equation (3.8) in the Laplace domain:

$$X_g(s) = (sI - A_g^d)^{-1} x_g(0) = ((sI - \Theta) - \Gamma)^{-1} x_g(0) \tag{3.14}$$

It is known that for any arbitrary square matrices Ω_1 and Ω_2:

$$(\Omega_1 + \Omega_2)^{-1} = \Omega_1^{-1} - \Omega_1^{-1} \left(I + \Omega_2 \Omega_1^{-1}\right)^{-1} \Omega_2 \Omega_1^{-1} \tag{3.15}$$

provided Ω_1 and $\left(I + \Omega_2\Omega_1^{-1}\right)$ are nonsingular. It can be concluded from (3.14) and (3.15) that:

$$X_g(s) = (sI - \Theta)^{-1} x_g(0) + (sI - \Theta)^{-1} \left(I - \Gamma(sI - \Theta)^{-1}\right)^{-1} \Gamma (sI - \Theta)^{-1} x_g(0) \tag{3.16}$$

Moreover, the assumption $x_0^{i,j} = x_i(0),\ i,j \in \bar{\nu},\ i \neq j$ yields that $x^i(0) = x(0),\ i \in \bar{\nu}$. Therefore,

$$x_g^i(0) = \left[\begin{array}{cc} x(0)^T & z^T(0) \end{array}\right]^T, \quad i \in \bar{\nu} \tag{3.17}$$

Hence, $x_g^i(0) = x_g^j(0),\ i,j \in \bar{\nu}$. From the equation (3.17) and the definitions of Θ and Γ given by (3.13), the ith entry of the vector $\Gamma(sI-\Theta)^{-1}x_g(0)$ can be obtained as follows:

$$\begin{aligned} &\tilde{A}_g^{i1}\left(sI-A_g^c\right)^{-1}x_g^1(0)+\cdots+\tilde{A}_g^{i(i-1)}\left(sI-A_g^c\right)^{-1}x_g^{i-1}(0)-\sum_{j=1,j\neq i}^{\nu}\tilde{A}_g^{ij}\left(sI-A_g^c\right)^{-1}x_g^i(0) \\ &+\tilde{A}_g^{i(i+1)}\left(sI-A_g^c\right)^{-1}x_g^{i+1}(0)+\cdots+\tilde{A}_g^{i\nu}\left(sI-A_g^c\right)^{-1}x_g^{\nu}(0)=0 \end{aligned} \tag{3.18}$$

It means that $\Gamma(sI-\Theta)^{-1}x_g(0) = 0$. Hence, it can be concluded from (3.16) that $X_g(s) = (sI-\Theta)^{-1}x_g(0)$. Consequently, from (3.17):

$$X_g^i(s) = \left(sI-A_g^c\right)^{-1}x_g^i(0) = \left(sI-A_g^c\right)^{-1}\left[\begin{array}{c} x(0) \\ z(0) \end{array}\right] \tag{3.19}$$

On the other hand, it follows from (3.10) that the state of the system $\mathscr{S}$ under the centralized controller K_c satisfies the following equation in the Laplace domain:

$$X(s) = \left(sI-A_g^c\right)^{-1}\left[\begin{array}{c} x(0) \\ z(0) \end{array}\right] \tag{3.20}$$

¿From (3.5) and (3.7), one can observe that the state of the subsystem S_i of the system $\mathscr{S}$ under the decentralized controller K_d is the ith block entry of $x_g^i(t)$ which is, according to (3.19), the ith block entry of $\left(sI-A_g^c\right)^{-1}\left[\begin{array}{c} x(0) \\ z(0) \end{array}\right]$ in the Laplace domain. On the other hand, equation (3.20) expresses that the state of the subsystem S_i of the system $\mathscr{S}$ under the centralized controller K_c is also the ith block entry of the same matrix. This means that the state of the system $\mathscr{S}$ under the decentralized and the centralized control configurations are equivalent. Using this result, it is straightforward now to show that the output and the input of these two closed-loop systems are the same. ∎

Theorem 1 states that under certain conditions, the state, the input, and the output of the system $\mathscr{S}$ under the centralized controller K_c are equal to those of the system $\mathscr{S}$ under the decentralized controller K_d. These conditions are met when the belief of each subsystem about the model and the initial state of any other subsystem is precise. Since these conditions are never met in practice, it is important to obtain stability conditions for the case when the corresponding beliefs are inaccurate. This issue is addressed in the following Corollary.

Corollary 1 *The interconnected system $\mathscr{S}$ under the proposed decentralized controller K_d is internally stable, if and only if all of the eigenvalues of the matrix A_g^d given in (3.9) are located in the left-half s-plane.*

Proof The proof follows immediately from the fact that the modes of the system $\mathscr{S}$ under the controller K_d are the eigenvalues of the matrix A_g^d, according to the equation (3.8). ■

Remark 1 *Note that stability of the decentralized closed-loop system is verified by simply checking the location of the eigenvalues of A_g^d. This signifies that it is independent of the values $x_0^{i,j}$, $i,j \in \bar{\nu}$, $i \neq j$.*

Remark 2 *The order of the local controller K_{d_i} for the ith subsystem of $\mathscr{S}$ is $n+\mu$ minus the order of the subsystem S_i, for any $i \in \bar{\nu}$. Moreover, it can be concluded from Theorem 1, that the local controller K_{d_i} implicitly includes an observer to estimate the states of the other subsystems. In contrast, if an explicit decentralized observer for each subsystem of $\mathscr{S}$ is desired to be designed by the existing methods, then the order of the observers $1,2,...,\nu$ (for subsystems $1,2,...,\nu$) will be $n,2n,...,n\nu$, respectively. Note that in a practical setup with explicit observers, each local controller consists of a compensator and an observer. This means that the local controllers designed here include implicit observers, whose orders are much less than those of conventional decentralized observers.*

3.4 Robustness analysis

In the previous section, a method was proposed for designing a decentralized controller based on a reference centralized controller and its stability condition was discussed in detail. Suppose that there are some uncertainties in the parameters of the system $\mathscr{S}$ given by (3.1). Since the decentralized controller K_d is designed in terms of the nominal parameters of the system, the resultant decentralized closed-loop system may become unstable. One of the main objectives of this section is to find a necessary and sufficient condition for internal stability of the perturbed decentralized closed-loop system.

Definition 5 *For any arbitrary matrix M, denote its perturbed version with $\bar{M}$, and definite its perturbation matrix as $\Delta M := \bar{M} - M$.*

Consider now the system $\bar{\mathscr{S}}$ as the perturbed version of the system $\mathscr{S}$ whose state-space representation is as follows:

$$\begin{aligned} \dot{x}(t) &= \bar{A}x(t) + \bar{B}u(t) \\ y(t) &= \bar{C}x(t) \end{aligned} \tag{3.21}$$

Let for any $i, j \in \bar{\mathrm{v}}$, the (i, j) block entry of the matrix $\bar{A}$ be denoted by $\bar{A}_{ij}$. Assume that the matrices $\bar{B}$ and $\bar{C}$ are block diagonal with the block entries $\bar{B}_1, \bar{B}_2, ..., \bar{B}_\mathrm{v}$ and $\bar{C}_1, \bar{C}_2, ..., \bar{C}_\mathrm{v}$, respectively. As discussed earlier, to construct the decentralized controller K_d, it is assumed that in addition to the output of any subsystem $\mathscr{S}_i$, all of the interconnections going out of subsystem i are available as the inputs for the local controller K_{d_i}. Hence, it is required to make some assumptions on the interconnection signals. Consider again the unperturbed system $\mathscr{S}$. Denote the interconnection signal coming out of subsystem i and going into subsystem j with $\zeta_{ji}(t)$. Since $\zeta_{ji}(t)$ can be considered as an output for subsystem i, there is a matrix Π_{ji} such that $\zeta_{ji}(t) = \Pi_{ji}x_i(t)$. Similarly, since $\zeta_{ji}(t)$ can be considered as an input for subsystem j, there is a matrix Γ_{ji} such that $A_{ji}x_j(t) = \Gamma_{ji}\zeta_{ji}(t)$. As a result, $A_{ji} = \Gamma_{ji}\Pi_{ji}$. Denote now the perturbed matrices corresponding to Π_{ji} and Γ_{ji} with $\bar{\Pi}_{ji}$ and $\bar{\Gamma}_{ji}$, respectively. Hence, $\bar{A}_{ji} = \bar{\Gamma}_{ji}\bar{\Pi}_{ji}$.

Definition 6 *1. Define $\bar{B}^i$ and $\bar{C}^i$, $i \in \bar{\mathrm{v}}$, as the matrices obtained from B^i and C^i by replacing their block entries B_i and C_i with $\bar{B}_i$ and $\bar{C}_i$, respectively.*

2. Define $\bar{\tilde{A}}_g^{ij}$, $i, j \in \bar{\mathrm{v}}$, $i \neq j$, as the perturbed matrix of $\tilde{A}_g^{ij}$, derived by replacing the block entry A_{ij} of $\tilde{A}_g^{ij}$ with $\bar{A}_{ij}$.

3. Define $\bar{A}^i$, $i \in \bar{\mathrm{v}}$, as the matrix derived from $\tilde{A}^i$ as follows:

- *Replace the entry A_{ii} with $\bar{A}_{ii}$.*
- *For all $j \in \bar{\mathrm{v}}$, $j \neq i$, replace the block entry A^i_{ji} with $\Gamma_{ji}\bar{\Pi}_{ji}$.*

Theorem 2 *Suppose that the decentralized controller K_d which is designed based on the nominal parameters of the system $\mathscr{S}$ as well as the centralized controller K_c, is applied to the perturbed system $\bar{\mathscr{S}}$. The resultant decentralized closed-loop control system is internally stable iff all of the* *eigenvalues of the matrix $\bar{A}_g^d$ are located in the open left-half of the complex plane, where*

$$\bar{A}_g^d := \begin{bmatrix} \bar{A}_g^1 & \bar{A}_g^{12} & \cdots & \bar{A}_g^{1\nu} \\ \bar{A}_g^{21} & \bar{A}_g^2 & \cdots & \bar{A}_g^{2\nu} \\ \vdots & \vdots & \ddots & \vdots \\ \bar{A}_g^{\nu 1} & \bar{A}_g^{\nu 2} & \cdots & \bar{A}_g^{\nu} \end{bmatrix} \tag{3.22}$$

and

$$\bar{A}_g^i := \begin{bmatrix} \bar{A}^i + \bar{B}^i D_k \bar{C}^i & \bar{B}^i C_k \\ B_k \bar{C}^i & A_k \end{bmatrix}, \quad i \in \bar{\nu} \tag{3.23}$$

Proof The proof of this theorem is omitted due to its similarity to Corollary 1. ■

Moreover, it can be easily verified that the modes of the perturbed system $\bar{\mathscr{S}}$ under the centralized controller K_c are the eigenvalues of the matrix $\bar{A}_g^c$, where

$$\bar{A}_g^c := \begin{bmatrix} \bar{A} + \bar{B} D_k \bar{C} & \bar{B} C_k \\ B_k \bar{C} & A_k \end{bmatrix} \tag{3.24}$$

Therefore, robustness analysis with respect to the perturbation in the parameters of the system can be summarized as follows:

- For decentralized case, the locations of the eigenvalues of the matrix $\bar{A}_g^d$ should be checked.
- For centralized case, the locations of the eigenvalues of the matrix $\bar{A}_g^c$ should be checked.

The equalities $A_{ij}^l = A_{ij}$, $B_i^l = B_i$, $C_i^l = C_i$, $i, j, l \in \bar{\nu}$, $i \neq l$ will hereafter be assumed to simplify the presentation of the properties of the decentralized control proposed in this chapter. It is to be noted that most of the results obtained under the above assumption, can simply be extended to the general case.

Theorem 3 *The Frobenius norms of the perturbation matrices for the decentralized and the centralized cases satisfy the following inequalities:*

$$\|\Delta(A_g^d)\| \leq \mathscr{N}_1 \tag{3.25a}$$

$$\|\Delta(A_g^c)\| \leq \mathscr{N}_2 \tag{3.25b}$$

where

$$(\mathscr{N}_1)^2 = 8\sum_{i=1}^{\nu} \|\Delta B_i\|^2\|D_k\|^2\|\Delta C_i\|^2 + 8\sum_{i=1}^{\nu} \|\Delta A_{ii}\|^2 + 4\sum_{\substack{i,j\in\bar{\nu}\\ i\neq j}} \|\Delta\Gamma_{ij}\|^2\|\Pi_{ij}\|^2$$
$$+\|\Delta B\|^2\|C_k\|^2 + 4\sum_{\substack{i,j\in\bar{\nu}\\ i\neq j}} \|\Delta\Gamma_{ij}\|^2\|\Delta\Pi_{ij}\|^2 + \|B_k\|^2\|\Delta C\|^2 + 12\sum_{\substack{i,j\in\bar{\nu}\\ i\neq j}} \|\Gamma_{ij}\|^2\|\Delta\Pi_{ij}\|^2 \quad (3.26)$$
$$+8\|B\|^2\|D_k\|^2\|\Delta C\|^2 + 8\|\Delta B\|^2\|D_k\|^2\|C\|^2$$

and

$$(\mathscr{N}_2)^2 = 32\sum_{\substack{i,j\in\bar{\nu}\\ i\neq j}} \|(\Delta\Gamma_{ij})\|^2\|\Pi_{ij}\|^2 + \|\Delta B\|^2\|C_k\|^2 + 32\sum_{\substack{i,j\in\bar{\nu}\\ i\neq j}} \|(\Delta\Gamma_{ij})\|^2\|\Delta\Pi_{ij}\|^2$$
$$+8\|\Delta B\|^2\|D_k\|^2\|\Delta C\|^2 + 8\|B\|^2\|D_k\|^2\|\Delta C\|^2 + 8\|\Delta B\|^2\|D_k\|^2\|C\|^2 \quad (3.27)$$
$$+32\sum_{\substack{i,j\in\bar{\nu}\\ i\neq j}} \|(\Gamma_{ij})\|^2\|\Delta\Pi_{ij}\|^2 + \|B_k\|^2\|\Delta C\| + 8\sum_{i=1}^{\nu} \|\Delta A_{ii}\|^2$$

and where $\|\cdot\|$ represents the Frobenius norm.

Proof One can write:

$$\|\Delta(A_g^d)\|^2 = \|\bar{A}_g^d - A_g^d\|^2 = \sum_{i,j\in\bar{\nu},\ i\neq j} \|\Delta A_{ij}\|^2 + \sum_{i=1}^{\nu} \|(\bar{B}^i - B)C_k\|^2 + \sum_{i=1}^{\nu} \|B_k(\bar{C}^i - C)\|^2$$
$$+\sum_{i=1}^{\nu} \|(\bar{A}^i - \tilde{A}^i) + (\bar{B}^i D_k \bar{C}^i - BD_kC)\|^2 \quad (3.28)$$

On the other hand,

$$\sum_{i=1}^{\nu} \left(\|(\bar{B}^i - B)C_k\|^2 + \|B_k(\bar{C}^i - C)\|^2\right) \leq \sum_{i=1}^{\nu} \left(\|(\bar{B}^i - B)\|^2\|C_k\|^2 + \|B_k\|^2\|(\bar{C}^i - C)\|^2\right)$$
$$\leq \sum_{i=1}^{\nu} \|\Delta B_i\|^2\|C_k\|^2 + \sum_{i=1}^{\nu} \|B_k\|^2\|\Delta C_i\|^2 \quad (3.29)$$
$$= \|\Delta B\|^2\|C_k\|^2 + \|B_k\|^2\|\Delta C\|^2$$

Furthermore,

$$\sum_{i,j\in\bar{\nu},\ i\neq j} \|\Delta A_{ij}\|^2 = \sum_{i,j\in\bar{\nu},\ i\neq j} \|\bar{\Gamma}_{ij}\bar{\Pi}_{ij} - \Gamma_{ij}\Pi_{ij}\|^2$$
$$= \sum_{i,j\in\bar{\nu},\ i\neq j} \|\Delta\Gamma_{ij}\Pi_{ij} + \Gamma_{ij}\Delta\Pi_{ij} + \Delta\Gamma_{ij}\Delta\Pi_{ij}\|^2$$
$$\leq 4\sum_{i,j\in\bar{\nu},\ i\neq j} \|\Delta\Gamma_{ij}\|^2\|\Pi_{ij}\|^2 + 4\sum_{i,j\in\bar{\nu},\ i\neq j} \|\Gamma_{ij}\|^2\|\Delta\Pi_{ij}\|^2 \quad (3.30)$$
$$+4\sum_{i,j\in\bar{\nu},\ i\neq j} \|\Delta\Gamma_{ij}\|^2\|\Delta\Pi_{ij}\|^2$$

and also,

$$\sum_{i=1}^{v} \|(\bar{A}^i - \tilde{A}^i) + (\bar{B}^i D_k \bar{C}^i - BD_k C)\|^2 \leq 8\sum_{i=1}^{v} \|(\bar{A}^i - \tilde{A}^i)\|^2 + 8\sum_{i=1}^{v} \|(\bar{B}^i - B)D_k(\bar{C}^i - C)\|^2$$
$$+8\sum_{i=1}^{v} \|(\bar{B}^i - B)D_k C\|^2 + 8\sum_{i=1}^{v} \|BD_k(\bar{C}^i - C)\|^2$$
$$\leq 8\sum_{i=1}^{v} \|\Delta A_{ii}\|^2 + 8 \sum_{i,j\in\bar{v},\ i\neq j} \|(\Gamma_{ij})\|^2 \|\bar{\Pi}_{ij} - \Pi_{ij}\|^2 \tag{3.31}$$
$$+8\sum_{i=1}^{v} \|\Delta B_i\|^2 \|D_k\|^2 \|\Delta C_i\|^2 + 8\sum_{i=1}^{v} \|B\|^2 \|D_k\|^2 \|\Delta C_i\|^2$$
$$+8\sum_{i=1}^{v} \|\Delta B_i\|^2 \|D_k\|^2 \|C\|^2$$

The above inequality is obtained by noting that the square of the summation of any four numbers is less than or equal to the summation of the squares of those numbers times eight. The proof of the inequality (3.25a) follows by substituting (3.29), (3.31) and (3.30) into (3.28). Pursuing a similar approach, one can easily obtain the inequality (3.25b). ■

The stability robustness analysis for the decentralized case can be described as follows. Consider a Hurwitz matrix A_g^d, and assume that it is desired to find admissible variations for the independent perturbation matrices ΔB_i, ΔC_i, $\Delta\Pi_{ij}$, $\Delta\Gamma_{ij}$, ΔA_{ii}, $i, j \in \bar{v}$, $i \neq j$ such that the perturbed matrix $\bar{A}_g^d$ remains Hurwitz. Analogously, for the centralized case one should check if $\bar{A}_g^c$, the perturbed version of the Hurwitz matrix A_g^c, is also Hurwitz. This kind of problem has been addressed in the literature using different approaches [11], [12].

To obtain some admissible variations for the independent perturbation matrices ΔB_i, ΔC_i, $\Delta\Pi_{ij}$, $\Delta\Gamma_{ij}$, ΔA_{ii}, $i, j \in \bar{v}$, $i \neq j$, one can choose any existing result for the matrix perturbation problem, e.g. Bauer-Fike Theorem, and substitute the bound $\mathcal{N}_1$ for the norm of the perturbation matrix to compute the admissible perturbations.

Remark 3 *Sensitivity of the eigenvalues of a matrix to the variation of its entries depends, in general, on several factors such as the structure of the matrix (represented by condition number or eigenvalue condition number [12]), repetition or distinction of the eigenvalues, and the most important of all, the norm of the perturbation matrix. On the other hand, it can be easily concluded from (3.26) and (3.27) that* the bound $\mathcal{N}_1$ for the decentralized case is less than or equal to the bound $\mathcal{N}_2$ for the centralized case. *Hence, it is expected that the decentralized closed-loop system be more robust to the parameter variation compared to the centralized closed-loop system.*

3.5 Performance evaluation

Since the perfect matching condition given in Theorem 1 does not hold in practice, it is desired now to evaluate the deviation in the performance of the decentralized closed-loop system with respect to its centralized counterpart. Define $\Delta x_i(t)$ as the state of the ith subsystem S_i of the closed-loop system $\mathscr{S}$ under K_d minus the state of the ith subsystem S_i of the closed-loop system $\mathscr{S}$ under K_c, for any $i \in \bar{\nu}$. Note that $\Delta x_i(t)$ is, in fact, the deviation in the state of the ith subsystem due to the mismatch between the real initial states and their expected values in the proposed decentralized control law.

To evaluate the closeness of the decentralized closed-loop system to the centralized closed-loop system, the following performance index is defined:

$$J_{dev} = \int_0^{\infty} \left(\Delta x(t)^T Q \Delta x(t) \right) dt \tag{3.32}$$

where

$$\Delta x(t) = \left[\begin{array}{cccc} \Delta x_1(t)^T & \Delta x_2(t)^T & \cdots & \Delta x_\nu(t)^T \end{array} \right]^T \tag{3.33}$$

and where $Q \in \Re^{n \times n}$ is a positive definite matrix. Consider now the system $\mathscr{S}$ under the decentralized controller K_d. This closed-loop system has the state $x_g(t)$, which consists of the states of the subsystems as well as those of the local controllers. However, only the states of the subsystems contribute to the performance index (3.32). Thus, it is desirable to derive $x(t)$ from $x_g(t)$. Define the following matrix:

$$\Phi_i = \left[\begin{array}{cccccccccc} 0_{n_i \times n_1} & 0_{n_i \times n_2} & \cdots & 0_{n_i \times n_{i-1}} & I_{n_i \times n_i} & 0_{n_i \times n_{i+1}} & \cdots & 0_{n_i \times n_\nu} & 0_{n_i \times \eta} \end{array} \right], \quad i \in \bar{\nu} \tag{3.34}$$

¿From the definition of $x_g^i(t)$ given in (3.7), one can write

$$x_i(t) = \Phi_i x_g^i(t), \quad i \in \bar{\nu} \tag{3.35}$$

Now, define the block diagonal matrix Φ as follows:

$$\Phi = \text{diag}\left([\Phi_1 \, , \, \Phi_2 \, , \, \ldots \, , \, \Phi_\nu] \right) \tag{3.36}$$

It can be concluded from (3.35) and the above matrix that $\Phi x_g(t) = x(t)$.

Definition 7 *Define Δx_0 as follows:*

$$\Delta x_0 = \left[\begin{array}{cccc} (\Delta x_0^1)^T & (\Delta x_0^2)^T & \cdots & (\Delta x_0^\nu)^T \end{array} \right]^T \tag{3.37}$$

where

$$\Delta x_0^i = \left[\begin{array}{cccccccc} (\Delta x_0^{1,i})^T & \cdots & (\Delta x_0^{i-1,i})^T & (0_{n_i \times 1})^T & (\Delta x_0^{i+1,i})^T & \cdots & (\Delta x_0^{\nu,i})^T & (0_{\eta \times 1})^T \end{array} \right]^T \tag{3.38}$$

for any $i \in \bar{\nu}$, and where

$$\Delta x_0^{i,j} = x_0^{i,j} - x_i(0), \quad i,j \in \bar{\nu}, \quad i \neq j \tag{3.39}$$

It is to be noted that Δx_0 can be considered as the prediction error of the initial states from different subsystems' view.

Theorem 4 *Suppose that the decentralized closed-loop system is internally stable. Then, the performance index J_{dev} given by (3.32) is equal to $\Delta x_0^T P_d \Delta x_0$, where the matrix P_d is the solution of the following Lyapunov equation:*

$$\left(A_g^d\right)^T P_d + P_d A_g^d + \Phi^T Q \Phi = 0 \tag{3.40}$$

Proof According to (3.8), the state of the interconnected system $\mathscr{S}$ under the proposed decentralized controller K_d satisfies the equation $\dot{x}_g(t) = A_g^d x_g(t)$. Also, Theorem 1 states that for the particular value of $x_0^{i,j} = x_i(0)$, the states of the subsystems of this decentralized closed-loop system are equal to the states of the subsystems of $\mathscr{S}$ under the centralized controller K_c. For this particular value, denote the state of the decentralized closed-loop system with $\bar{x}_g(t)$, which satisfies the equation $\dot{\bar{x}}_g(t) = A_g^d \bar{x}_g(t)$ as well. Subtracting these two equations and using Definition 4, results in:

$$\Delta \dot{x}_g(t) = A_g^d \Delta x_g(t), \quad \Delta x_g(0) = \Delta x_0 \tag{3.41}$$

where $\Delta x_g(t) = \bar{x}_g(t) - x_g(t)$. Therefore:

$$\begin{aligned} J_{dev} &= \int_0^\infty \Delta x(t)^T Q \Delta x(t) dt = \int_0^\infty \left(\Phi \Delta x_g(t)\right)^T Q \left(\Phi \Delta x_g(t)\right) dt \\ &= \int_0^\infty \Delta x_g(t)^T \Phi^T Q \Phi \Delta x_g(t) dt \end{aligned} \tag{3.42}$$

Thus, it can be concluded from (3.40), (3.41) and (3.42) that:

$$\begin{aligned}
J_{dev} &= -\int_0^\infty \Delta x_g(t)^T \left(\left(A_g^d\right)^T P_d + P_d A_g^d \right) \Delta x_g(t) dt \\
&= -\int_0^\infty \left(\left(A_g^d \Delta x_g(t)\right)^T P_d \Delta x_g(t) + \Delta x_g(t)^T P_d \left(A_g^d \Delta x_g(t)\right) \right) dt \\
&= -\int_0^\infty \left(\Delta \dot{x}_g(t)^T P_d \Delta x_g(t) + \Delta x_g(t)^T P_d \Delta \dot{x}_g(t) \right) dt \\
&= -\int_0^\infty \frac{d\left(\Delta x_g(t)^T P_d \Delta x_g(t)\right)}{dt} dt
\end{aligned} \tag{3.43}$$

On the other hand, stability of the decentralized closed-loop system implies that $\Delta x_g(\infty) = 0$. Furthermore, since $\Delta x_g(0) = \Delta x_0$, one can rewrite (3.43) as follows:

$$J_{dev} = -\Delta x_g(\infty)^T P_d \Delta x_g(\infty) + \Delta x_g(0)^T P_d \Delta x_g(0) = \Delta x_0^T P_d \Delta x_0 \tag{3.44}$$

■

Consider now the interconnected system $\mathscr{S}$ under the centralized control law K_c, and define the following performance index for it:

$$J_c = \int_0^\infty x(t)^T Q x(t) dt \tag{3.45}$$

It is straightforward to use a similar approach and apply it to (3.10) to show that $J_c = x(0)^T P_c x(0)$, where the matrix P_c is the solution of the following Lyapunov equation:

$$\left(A_g^c\right)^T P_c + P_d A_g^c + Q = 0 \tag{3.46}$$

Remark 4 *One can use Theorem 4 and the equation (3.46) to obtain statistical results for the relative performance deviation* $\frac{J_{dev}}{J_c}$ *in terms of the expected values of the initial states of the subsystems. This can be achieved by using Chebyshev's inequality. This enables the designer to determine the maximum allowable standard deviation for* Δx_0 *to achieve a relative performance deviation within a prespecified region with a sufficiently high probability (e.g. 95%).*

3.6 Near-optimal decentralized control law

Consider the interconnected system $\mathscr{S}$ given by (3.1), and suppose that it is desired to find the controller K_c given by (3.3) in order to minimize the following performance index:

$$J = \int_0^\infty \left(x(t)^T Q x(t) + u(t)^T R u(t) \right) dt \tag{3.47}$$

where $R \in \Re^{m\times m}$ and $Q \in \Re^{n\times n}$ are positive definite and positive semi-definite matrices, respectively. Without loss of generality, assume that Q and R are symmetric. If C is an identity matrix (i.e., if all states are available in the output), then the solution of this optimization problem is as follows:

$$u(t) = -K_{opt}y(t) = -K_{opt}x(t) \tag{3.48}$$

where the matrix K_{opt} is obtained from the Riccati equation. Assume now that it is desired to design a decentralized controller, instead of the centralized controller (3.48), such that the performance index (3.47) is minimized, under assumption that $C = I_{n\times n}$. Unlike the centralized case, computation of the optimal decentralized controller can be cumbersome. Therefore, many of the existing results, instead, present a near-optimal decentralized controller (instead of an optimal one) with a static (local output or local state) feedback structure. It is clear that this constraint can significantly affect the performance of the decentralized closed-loop system. In this section, the proposed method for designing a decentralized controller is exploited to present a near-optimal decentralized control law whose performance will later be evaluated.

Assume that the controller K_c given by (3.3) is the optimal centralized controller given by (3.48), i.e.

$$A_k = 0, \quad ,B_k = 0, \quad C_k = 0, \quad D_k = -K_{opt} \tag{3.49}$$

Construct the proposed decentralized controller K_d based on this controller K_c as described in Section 3.3. It was shown that this decentralized controller relies on some constant values $x_0^{i,j}$, $i,j \in \bar{\nu}$, $i \neq j$, and in a particular case when $x_0^{i,j} = x_i(0)$, $i,j \in \bar{\nu}$, $i \neq j$, the system $\mathscr{S}$ under K_d behaves exactly the same as the system $\mathscr{S}$ under K_c. Denote the values of the performance index (3.47) for the centralized and decentralized cases with J_c and J_d, respectively. Let the performance deviation be defined as $\Delta J := J_d - J_c$. Also, suppose that $x(t)$ and $u(t)$ are the state and the input of the centralized control system excluding the state and the input of the controller, while $x(t) + \Delta x(t)$ and $u(t) + \Delta u(t)$ denote those of the decentralized case.

Theorem 5 *ΔJ satisfies the following equation:*

$$\Delta J = \int_0^{\infty} \left(\Delta x^T Q \Delta x + \Delta u^T R \Delta u\right) dt \tag{3.50}$$

Proof Using an approach similar to the proof of Theorem 4, one can rewrite J_d as:

$$J_d = \begin{bmatrix} x(0)^T & \Delta x_0^T \end{bmatrix} \begin{bmatrix} V_{11} & V_{12} \\ V_{12}^T & V_{22} \end{bmatrix} \begin{bmatrix} x(0) \\ \Delta x_0 \end{bmatrix} \tag{3.51}$$

where the matrices $V_{11} \in \Re^{n\times n}, V_{12} \in \Re^{n\times(n+\eta)\nu}$ and $V_{22} \in \Re^{(n+\eta)\nu\times(n+\eta)\nu}$ are functions of A, B and K_{opt}. The fact that the $(2,1)$ block entry of the above matrix is transpose of its $(1,2)$ block entry V_{12}, results from the symmetry of the Q and R matrices. According to Theorem 1, when Δx_0 is zero, the centralized and decentralized closed-loop systems are identical. Therefore, by substituting $\Delta x_0 = 0$ into the equation (3.51), J_c will be obtained as follows:

$$J_c = x(0)^T V_{11} x(0) \tag{3.52}$$

It can be concluded from (3.51) and (3.52) that

$$\Delta J = \begin{bmatrix} x(0)^T & \Delta x_0^T \end{bmatrix} \begin{bmatrix} 0 & V_{12} \\ V_{12}^T & V_{22} \end{bmatrix} \begin{bmatrix} x(0) \\ \Delta x_0 \end{bmatrix} \tag{3.53}$$

Suppose now that $x(0)$ is any arbitrary vector. It is desired to find a simple closed from relationship between the performance deviation ΔJ and the initial state prediction error Δx_0. ΔJ given in (3.53) has the following properties:

- ΔJ is always nonnegative, because the centralized optimal performance index has the smallest value among all the performance indices obtained by using any type of control.
- Substituting $\Delta x_0 = 0$ in (3.53) yields $\Delta J = 0$.
- ΔJ is continuous with respect to each of the entries of Δx_0, because ΔJ is quadratic.

It can be concluded from the above properties that $\Delta x_0 = 0$ is an extremum point for ΔJ. Thus, the partial derivative of ΔJ with respect to Δx_0 has to be zero at $\Delta x_0 = 0$. Hence:

$$\left[x(0)^T V_{12} + \left(V_{12}^T x(0)\right)^T + \Delta x_0^T \left(V_{22}^T + V_{22}\right)\right]\Bigg|_{\Delta x_0 = 0} = 0 \tag{3.54}$$

Accordingly, $x(0)^T V_{12} = 0$. This implies that $x(0)$ is in the null-space of the matrix V_{12}. Since $x(0)$ is any arbitrary vector, thus $V_{12} = 0$. Substituting this result into (3.53), it can be concluded that $\Delta J = \Delta x_0 V_{22} \Delta x_0$. This means that ΔJ does not directly depend on $x(0)$. Furthermore,

$$\begin{aligned} \Delta J = & \int_0^\infty \left([x+\Delta x]^T Q[x+\Delta x] + [u+\Delta u]^T R[u+\Delta u]\right) dt - \int_0^\infty \left(x^T Qx + u^T Ru\right) dt \\ = & \int_0^\infty \left(\Delta x^T Q\Delta x + \Delta u^T R\Delta u\right) dt + \int_0^\infty \left(x^T Q\Delta x + u^T R\Delta u\right) dt \\ + & \int_0^\infty \left(\Delta x^T Qx + \Delta u^T Ru\right) dt \end{aligned} \tag{3.55}$$

Using the equation (3.48) and after some mathematical computations, one can easily conclude that the summation of the second and the third integrals in the right side of the last equation above is in the form of $x(0)^T \bar{V} \Delta x_0 + \Delta x_0^T \bar{V}^T x(0)$, respectively. Since it was shown that ΔJ is not directly dependent on $x(0)$, the summation of these two integrals should be zero. This completes the proof. ■

Definition 8 *Define the block diagonal matrix $\bar{\Phi}$ as follows:*

$$\bar{\Phi} = \text{diag}\left([K_{opt} \ , \ 0_{\eta\times\eta} \ , \ K_{opt} \ , \ 0_{\eta\times\eta} \ , \ \cdots \ , \ K_{opt} \ , \ 0_{\eta\times\eta}] \right) \tag{3.56}$$

where the block entry K_{opt} appears ν times in the above matrix.

Since $u^i(t) = -K_{opt} x^i(t)$, it is straightforward to verify that the vector $\Phi\bar{\Phi}x_g^d(t)$ (the matrix Φ is given by (3.36)) is equal to the input of the system $\mathscr{S}$ under K_d, which is denoted by $u(t) + \Delta u(t)$.

Theorem 6 *ΔJ can be written as $\Delta x_0^T \bar{P} \Delta x_0$, where $\bar{P}$ is the solution of the following Lyapunov equation:*

$$\left(A_g^d\right)^T \bar{P} + \bar{P} A_g^d + (\Phi^T Q \Phi + \bar{\Phi}^T \Phi^T R \Phi \bar{\Phi}) = 0 \tag{3.57}$$

Proof The proof is omitted due to its similarity to the proof of Theorem 4. ■

3.7 Numerical example

Consider an interconnected system consisting of two SISO subsystems, with the following parameters:

$$A = \begin{bmatrix} 1 & -2 \\ 1 & 2 \end{bmatrix}, \quad B = \begin{bmatrix} -1 & 0 \\ 0 & 1 \end{bmatrix}, \quad C = I_{2\times 2} \tag{3.58}$$

Suppose that a centralized controller K_c is given for this system with the following parameters:

$$A_k = \begin{bmatrix} -1 & -2 \\ 1 & -1 \end{bmatrix}, \quad B_k = \begin{bmatrix} 1 & 1 \\ 2 & 1 \end{bmatrix}, \quad C_k = I_{2\times 2}, \quad D_k = \begin{bmatrix} 1 & 2 \\ 1 & -3 \end{bmatrix} \tag{3.59}$$

It is desired now to design a decentralized controller K_d, such that the system under K_d behaves as closely as possible to the system under K_c. Using the method proposed in this chapter, K_d can be designed in terms of two constant values $x_0^{1,2}$ and $x_0^{2,1}$, which are the expected values of each subsystem's initial state from the other subsystem's view. For simplicity, suppose that $x_1(0) = x_2(0) = 1$, $x_k(0) = \begin{bmatrix} 1 & 1 \end{bmatrix}^T$. Consider now the following two mismatching cases:

i) Suppose that $x_0^{2,1} = 0.5$ (-50% prediction error) and $x_0^{1,2} = 1.5$ (50% prediction error). Note that the numbers within the above parentheses show the percentage of errors in predicting the initial states. The output of the first subsystem is sketched with both centralized and decentralized controllers, in Figure 3.1 (note that the output of the second subsystem is not depicted here due to space restrictions). It is evident that the output trajectory of the decentralized case is very close to that of the centralized case.

ii) Assume that $x_0^{2,1} = 0$ (-100% prediction error) and $x_0^{1,2} = 2$ (100% prediction error). Figure 3.2 illustrates the output of the first subsystem, with both centralized and decentralized controllers. Despite the large prediction errors, the outputs are close to each other.

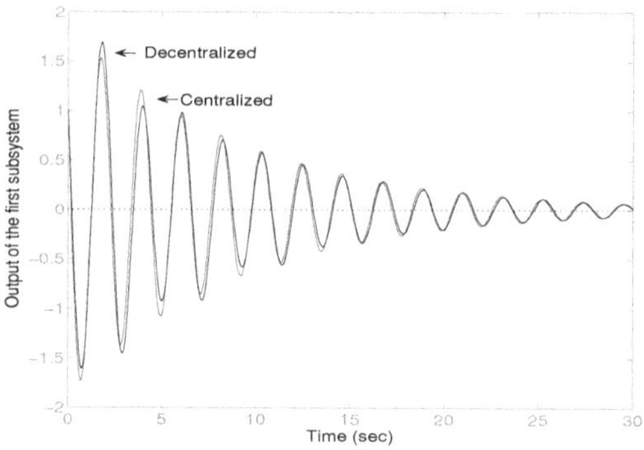

Figure 3.1: The output of the first subsystem with both centralized and decentralized controllers. The prediction errors of the initial states are $\pm 50\%$.

Now, to evaluate the performance of the proposed decentralized controller with respect to its centralized counterpart, consider the performance index defined in (3.32) and assume that $Q = I$. It can be concluded from Theorem 4 that:

$$J_{dev} = 1.128\left(x_0^{2,1}\right)^2 + 1.107\left(x_0^{1,2}\right)^2 - 0.259 x_0^{2,1} x_0^{1,2} \tag{3.60}$$

On the other hand, it can be easily verified that $J_c = 18.147$ by using the equation (3.46). Now, one can find that the expected value and the standard deviation of $\frac{J_{dev}}{J_c}$, provided some probabilistic data regarding $x_0^{1,2}$ and $x_0^{2,1}$ are available.

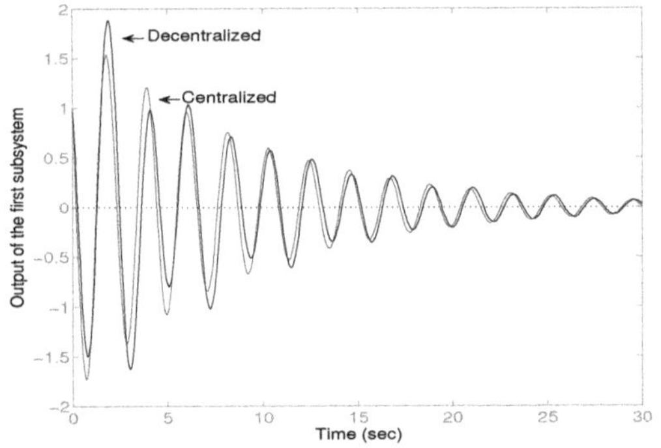

Figure 3.2: The output of the first subsystem with both centralized and decentralized controllers. The prediction errors of the initial states are $\pm 100\%$.

3.8 Conclusions

A decentralized controller is proposed for interconnected systems. The control law is constructed based on the parameters of a given centralized controller and *a priori* knowledge about each subsystem's model from any other subsystem's view. It is shown that the proposed controller behaves exactly the same as its centralized counterpart, provided the knowledge of each subsystem has no error. Furthermore, a set of conditions for the stability of the decentralized closed-loop system in presence of inexact knowledge of the subsystems' model as well as perturbation in the system's parameters is presented. Moreover, it is shown that the decentralized control system is likely more robust than the centralized control system. In addition, a quantitative measure is given to statistically assess the closeness of the resultant decentralized closed-loop performance to its centralized counterpart. The proposed method is also used to design a near-optimal decentralized control law. The deviation of the decentralized near-optimal performance index from the centralized optimal performance index is obtained in a closed form, which enables the designer to determine how small the standard deviation of the initial state predictions should be in order to achieve a prespecified performance index with a certain likelihood. Simulation results demonstrate the effectiveness of the proposed decentralized control law.

Bibliography

[1] E. J. Davison and T. N. Chang, "Decentralized stabilization and pole assignment for general proper systems," *IEEE Trans. Automat. Contr.*, vol. 35, no. 6, pp. 652-664, June 1990.

[2] N. R. Sandell, P. Varaiya, M. Athans and M. G. Safonov, "Survey of decentralized control methods for large scale systems," *IEEE Trans. Automat. Contr.*, vol. 23, no. 2, pp. 108-128, Apr. 1978.

[3] D.D. Šiljak, Decentralized control of large complex systems, Boston: Academic Press, 1991.

[4] H. Cuui and E. W. Jacobsen, "Performance limitations in decentralized control," *J. Process Contr.*, vol. 12, no. 4, pp. 485-495, 2002.

[5] R. Krtolica and D. D. Šiljak, "Suboptimality of decentralized stochastic control and estimation," *IEEE Trans. Automat. Contr.*, vol. 25, no. 1, pp. 76-83, Feb. 1980.

[6] K. D. Young, "On near optimal decentralized control," *Automatica*, vol. 21, no. 5, pp. 607-610, Sep. 1985.

[7] H. T. Toivoneh and P.M. Makila, " A descent Anderson-Moore algorithm for optimal decentralized control," *Automatica*, vol. 21, no. 6, pp. 743-744, Nov. 1985.

[8] S. V. Savastuk and D. D. Šiljak, "Optimal decentralized control," *in Proc. 1994 American Contr. Conf.*, Baltimore, MD, vol. 3, pp. 3369-3373, 1994.

[9] D. D. Sourlas and V. Manousiouthakis, "Best achievable decentralized performance," *IEEE Trans. Automat. Contr.*, vol. 40, no. 11, pp. 1858-1871, Nov. 1995.

[10] D. M. Stipanovi, G. Inalhan, R. Teo and C. J. Tomlin, "Decentralized overlapping control of a formation of unmanned aerial vehicles," *Automatica*, vol. 40 , no.8, pp. 1285-1296, Aug. 2004.

[11] M. Embree and L. N. Trefethen, "Generalizing eigenvalue theorems to pseudospectra theorems," *SIAM I. SCI. COMPUT.*, vol. 23, no. 2, pp. 583-590, 2001.

[12] S. C. Eisenstat and I. C. F. Ipsen, "Three absolute perturbation bounds for matrix eigenvalues imply relative bounds," *SIAM J. Matrix Anal. Appl*, vol. 20, no. 1, pp. 149-158, 1998.

Chapter 4

Structurally Constrained Periodic Feedback Design

4.1 Abstract

This chapter aims to design a high-performance controller with any predefined structure for continuous-time LTI systems. The control law employed is generalized sampled-data hold fucntion (GSHF), which can have any special form, e.g. polynomial, exponential, piecewise constant, etc. The GSHF is first written as a linear combination of a set of basis functions obtained in accordance with its desired form and structure. The objective is to find the coefficients of this linear combination, such that a prespecified linear-quadratic performance index is minimized. A necessary and sufficient condition for the existence of such GSHF is first obtained in the form of matrix inequality, which can be solved by using the existing methods to obtain a set of stabilizing initial values for the coefficients or to conclude the non-existence of such structurally constrained GSHF. An efficient algorithm is then presented to compute the optimal coefficients from their initial values, so that the performance index is minimized. This chapter utilizes the latest developments in the area of sum-of-square polynomials. The effectiveness of the proposed method is demonstrated in two numerical examples.

4.2 Introduction

There has been a considerable amount of interest in the past several years towards control of continuous-time systems by means of periodic feedback, or so-called generalized sampled-data hold functions (GSHF) [1; 2; 3; 4]. Periodic feedback control signal is constructed by sampling the output of the system at equidistant time instants, and multiplying the samples by a continuous-time hold function, which is defined over one sampling interval. Several advantages and disadvantages of GSHF and its application in practical problems have been thoroughly investigated in the literature and different design techniques are proposed [3; 4; 5].

It is known that if a system has an unstable fixed mode with respect to a given information flow structure, there is no LTI controller to stabilize the system [6]. In that case, under some conditions a time-varying control law, e.g. periodic feedback, can be used to stabilize the system [7; 8]. It is shown in [9] that using GSHF with exponential form may eliminate the unstable fixed modes. However, designing a controller which only takes stability into account is not beneficial in practice, as it may not provide a satisfactory performance. This issue is addressed to some extent in [10] by minimizing a performance index. Nevertheless, only a piecewise-constant GSHF is considered there.

In this chapter, the problem of structurally constrained optimal GSHF with respect to a quadratic continuous-time performance index is considered. The main objective is to design a GSHF which satisfies the following constraints:

i) It stabilizes the plant.

ii) It has the desired decentralized structure.

iii) It has a prespecified form such as polynomial, piecewise constant, etc.

iv) It minimizes a predefined guaranteed cost function.

It is to be noted that condition (iii) given above is motivated by the following practical issues:

- In many problems involving robustness, noise rejection, simplicity of implementation, elimination of fixed modes, etc., it is desired to design GSHFs with a specific form, e.g. piecewise constant, exponential, etc. [11; 9; 3].

- Design of a high performance stabilizing *piecewise constant* GSHF is studied in [10], which can be classified as a special case of the most general form considered in the present chapter.

- Design of unconstrained optimal GSHF using a continuous-time quadratic performance index has been studied by several researchers [12; 13]. The optimal GSHF is derived from a two-boundary point partial differential equation, which is very difficult to solve analytically [12]. Different methods are proposed to solve the problem numerically [13; 14]. These methods are iterative and may not be computationally efficient in general.

In this chapter, conditions (ii) and (iii) are formulated by writing the GSHF as a linear combination of appropriate basis functions. The problem is then reduced to finding the coefficients of the linear combination, such that conditions (i) and (iv) are met. It is shown that the aforementioned problem is solvable (i.e., the desired GSHF exists), if and only if another system which is derived from the original one is stabilizable by means of a constrained *static* output feedback. A method is then proposed to solve the optimization problem for each coefficient analytically, by keeping all remaining coefficients unchanged. The coefficients are obtained one at a time and can be improved by solving the optimization problem with respect to each coefficient iteratively, using the new improved values for other coefficients at each iteration.

4.3 Existence of a stabilizing constrained GSHF

Consider a system $\mathscr{S}$ with the following state-space representation:

$$\dot{x}(t) = Ax(t) + Bu(t) \tag{4.1a}$$

$$y(t) = Cx(t) \tag{4.1b}$$

where $x \in \Re^n$, $u \in \Re^m$ and $y \in \Re^r$ are the state, the input and the output, respectively, and the matrices A, B and C have proper dimensions. It is desired to design a GSHF $f(t)$ which minimizes the following performance index:

$$J = \int_0^{\infty} \left(x(t)^T Q x(t) + u(t)^T R u(t) \right) dt \tag{4.2}$$

where $R \in \Re^{m \times m}$ and $Q \in \Re^{n \times n}$ are positive definite and positive semidefinite matrices, respectively. It is to be noted that since (4.2) is a continuous-time performance index, it takes the intersample ripple effect into account. Consider now a set of basis functions for $f(t)$, denoted by:

$$\mathbf{f} := \{f_1(t), f_2(t), ..., f_k(t)\} \tag{4.3}$$

where:

$$f_i(t) = f_i(t+h), \quad i = 1,2,...,k, \; t \geq 0 \tag{4.4}$$

and where $f_i(t) \in \Re^{m \times r}$, $i = 1,2,...,k$, are matrices with only one nonzero element each (this is clarified in Examples 1 and 2 of Section 4.5). It is to be noted that the equation (4.4) implies that the basis functions are periodic with the period h. Assume that $f(t)$ is desired to be in the form of a linear combination of the basis functions in $\mathbf{f}$, as follows:

$$f(t) = \alpha_1 f_1(t) + \alpha_2 f_2(t) + \cdots + \alpha_k f_k(t) \tag{4.5}$$

It is to be noted that the set of basis functions (4.3) can be easily found, once the desired form (e.g. polynomial, piecewise constant, etc.) and structure (e.g. block diagonal, etc.) of the GSHF is specified. The desired structure is determined based on the information flow matrix which represents the control constraint. The motivation for using a specific form for GSHF, on the other hand, was discussed in Section 4.2. For instance, one can use a GSHF of the polynomial form with any arbitrary order, as an approximation to the Taylor series of the optimal GSHF, which is very difficult to find in a closed form, in general.

The input $u(t)$ is related to the samples of the output through the following equation:

$$u(t) = f(t)y[\kappa], \quad \kappa h \leq t < (\kappa+1)h, \quad \kappa \geq 0 \tag{4.6}$$

Note that the discrete argument corresponding to the samples of any signal is enclosed in brackets (e.g., $y[\kappa] := y(\kappa h)$). It is known that the state of the system (4.1) under the control law (4.6) is given by:

$$x(t) = e^{(t-\kappa h)A}x(\kappa h) + \int_{\kappa h}^{t} e^{(t-\tau)A}Bu(\tau)d\tau \tag{4.7}$$

for any $\kappa h \leq t \leq (\kappa+1)h$, $\kappa \geq 0$. Let the following matrices be defined:

$$M_0(t) := e^{tA}, \tag{4.8a}$$

$$M_i(t) := \int_0^t e^{(t-\tau)A}Bf_i(\tau)Cd\tau, \quad i = 1,2,...,k \tag{4.8b}$$

$$M(t,\alpha) := M_0(t) + \sum_{i=1}^{k} \alpha_i M_i(t) \tag{4.8c}$$

where $\alpha := [\alpha_1, \alpha_2, ..., \alpha_k]$. One can easily conclude from (4.1b), (4.6), (4.7), and (4.8), that:

$$x(t) = M(t - \kappa h, \alpha)x[\kappa], \quad \kappa h \leq t \leq (\kappa + 1)h \tag{4.9}$$

Consequently:

$$x[\kappa] = (M(h, \alpha))^{\kappa} x[0], \quad \kappa = 0, 1, 2, ... \tag{4.10}$$

Now, define:

$$\begin{aligned} N(\alpha) &:= \int_0^h M^T(t, \alpha) Q M(t, \alpha) dt + \int_0^h C^T f(t)^T R f(t) C dt \\ \tilde{M}(t) &:= \begin{bmatrix} M_1(t)^T & M_2(t)^T & \cdots & M_k(t)^T \end{bmatrix}^T \end{aligned} \tag{4.11}$$

Lemma 1 *There exists a GSHF $f(t)$ with the desired structure (given by the equation (4.5)) such that the system $\mathscr{S}$ is stable under the periodic feedback law (4.6) if and only if there exist a symmetric positive definite matrix Ω, and scalars $\alpha_1, \alpha_2, ..., \alpha_k$ such that the matrix $\tilde{\alpha} := \begin{bmatrix} \alpha_1 I_n & \alpha_2 I_n & \cdots & \alpha_k I_n \end{bmatrix}$ satisfies the following inequality:*

$$\left(M_0(h) + \tilde{\alpha}\tilde{M}(h)\right)^T \Omega \left(M_0(h) + \tilde{\alpha}\tilde{M}(h)\right) - \Omega + (\tilde{\alpha}\tilde{M}(h))^T (\tilde{\alpha}\tilde{M}(h)) < 0 \tag{4.12}$$

Proof It can be concluded from (4.9) that the system $\mathscr{S}$ is stable under the feedback law (4.6) if and only if $x[k] \to 0$ as $k \to \infty$. On the other hand, $x[k] \to 0$ as $k \to \infty$ if and only if all of the eigenvalues of the matrix $M(h, \alpha)$ are located inside the unit circle in the complex plane (according to (4.10)). As a result, the system $\mathscr{S}$ is stable under the periodic feedback law (4.6) if and only if all of the eigenvalues of the matrix $M(h, \alpha) = M_0(h) + \tilde{\alpha}\tilde{M}(h)$ are located inside the unit circle, or equivalently, the system with the representation:

$$\begin{aligned} \bar{x}[\kappa + 1] &= M_0(h)\bar{x}[\kappa] + I_n \bar{u}[\kappa] \\ \bar{y}[\kappa] &= \tilde{M}(h)\bar{x}[\kappa] \end{aligned} \tag{4.13}$$

is stabilizable by a static output feedback with the gain $\tilde{\alpha}$. This problem is usually referred to as "stabilization of a LTI system via structured static output feedback", which has been investigated intensively in the literature; e.g. see [10] and the references therein. The proof follows immediately by applying the necessary and sufficient condition for the existence of a stabilizing static output feedback gain obtained in [10] to the system given by (4.13). ■

Remark 1 *It results from the proof of Lemma 1, that for any given α, the system $\mathscr{S}$ is stable under the feedback law (4.6) iff all of the eigenvalues of the matrix $M(h,\alpha)$ are located inside the unit circle in the complex plane.*

Lemma 1 presents a necessary and sufficient condition for the existence of a structurally constrained GSHF $f(t)$ which stabilizes the system $\mathscr{S}$. One can exploit the LMI algorithm proposed in [10] to solve the matrix inequality (4.12) in order to obtain initial stabilizing values for the matrices α_i, $i = 1,2,...,k$, denoted by θ_i, $i = 1,2,...,k$, or conclude the non-existence of such GSHF $f(t)$.

Lemma 2 *Suppose that the unknown coefficients $\alpha_1, \alpha_2, ..., \alpha_k$ are such that the GSHF $f(t)$ stabilizes the system $\mathscr{S}$. The performance index J can be written as:*

$$J = x^T(0)K(\alpha)x(0) \tag{4.14}$$

where $K(\alpha)$ satisfies the following discrete Lyapunov equation:

$$M^T(h,\alpha)K(\alpha)M(h,\alpha) - K(\alpha) + N(\alpha) = 0 \tag{4.15}$$

Proof Substituting (4.9) and (4.6) into (4.2) and using (4.10) and (4.11), the performance index can be rewritten as follows:

$$\begin{aligned} J &= \sum_{\kappa=0}^{\infty}\left(\int_{\kappa h}^{(\kappa+1)h}\left(x^T(t)Qx(t) + u^T(t)Ru(t)\right)dt\right) = \sum_{\kappa=0}^{\infty}\left(x^T[\kappa]N(\alpha)x[\kappa]\right) \\ &= x^T(0)\sum_{\kappa=0}^{\infty}\left(M^T(h,\alpha)\right)^{\kappa}N(\alpha)\left(M(h,\alpha)\right)^{\kappa}x(0) \end{aligned} \tag{4.16}$$

Since the system $\mathscr{S}$ is assumed to be stable under the feedback law (4.6), the infinite series given above is converging. The proof follows from the well-known property that the solution of this series satisfies the discrete Lyapunov equation given by (4.15), if the series is convergent. ■

As a result of Lemma 2 and Remark 1, the following steps should be taken in order to find the optimal values for $\alpha_1, \alpha_2, ..., \alpha_k$ which minimize the performance index (4.2):

1. *Finding a region of convergence (ROC) for the infinite series in (4.16):* As discussed earlier, convergence of the series in (4.16) follows from the condition that all of the eigenvalues of the matrix $M(h,\alpha)$ lie inside the unit circle. Since there are k variables $\alpha_1, \alpha_2, ..., \alpha_k$, the corresponding ROC is a region in the k-dimensional space. It is important to note that this ROC is

not empty, because it contains at least the following point (and a sufficiently close neighborhood of it, due to continuity):

$$(\alpha_1, \alpha_2, ..., \alpha_k) = (\theta_1, \theta_2, ..., \theta_k)$$

It is to be noted that the non-emptiness of the ROC is required here because the optimization will later be carried out in this region. Finding this ROC, in general, can be cumbersome. However, using some inequalities, a conservative ROC can be obtained (for example by using Bauer-Fike theorem, which will be explained later).

2. *Solving the discrete Lyapunov equation given in (4.15):* Since this equation is parametric, it cannot be solved easily by any computer software.

3. *Solving a constrained optimization problem:* After finding the ROC (or a subset of it) and solving the Lyapunov equation and substituting its solution into (4.14), the performance index will be obtained in terms of $\alpha_1, \alpha_2, ..., \alpha_k$, which is valid in the obtained ROC. Now, the performance index function should be minimized over this region. This is a constrained global optimization problem, which is not solvable in general, if the problem is not convex.

In the next section, a remedy for the drawbacks of the above steps will be presented.

4.4 Optimal performance index

To optimize a multivariate function numerically, when its derivative or gradient is unavailable, one can use the following procedure, which is usually used in direct search methods [15]:
Consider all except one of the variables in the optimization problem as constants. Then, try to find the optimal value for that particular variable. Similarly, select another variable, set all other variables to be fixed, and solve the optimization problem with respect to that variable. Continue this procedure for all variables one by one (it may require to repeat several times).

It is to be noted that the above procedure may lead to a locally optimal point. This method will now be used to solve the optimization problem formulated in the previous section, and its effectiveness will later be discussed. Since only one variable is considered in each step of the procedure and all other variables are set to be fixed, the problem can be reformulated as follows:

Suppose that $\bar{f}(t)$ is a stabilizing GSHF for the system (4.1) in a closed-loop set-up. Suppose also that it is desired to add a term proportional to a given function $g(t)$ in the set of basis functions **f** *given by (4.3) to generate a new GSHF $\bar{f}(t)$, i.e.:*

$$f(t) = \bar{f}(t) + \alpha g(t) \tag{4.17}$$

where α is a real constant number, which is desired to be found such that the continuous-time performance index (4.2) is minimized. Note that the GSHF $f(t)$ stabilizes the system $\mathscr{S}$ for $\alpha = 0$. Define the following matrices similar to (4.8):

$$\begin{aligned} M_0(t) &:= e^{tA} + \int_0^t e^{(t-\tau)A} B\bar{f}(\tau) C d\tau, \\ M_1(t) &:= \int_0^t e^{(t-\tau)A} Bg(\tau) C d\tau, \\ M(t,\alpha) &:= M_0(t) + \alpha M_1(t) \end{aligned}$$

Consider now the matrix defined in (4.11). It can be written as follows:

$$N(\alpha) := P_0 + P_1\alpha + P_2\alpha^2$$

where:

$$\begin{aligned} P_0 &:= \int_0^h \left(M_0(t)^T Q M_0(t) + C^T \bar{f}(t)^T R \bar{f}(t) C\right) dt \\ P_1 &:= \int_0^h \left(M_0(t)^T Q M_1(t) + M_1(t)^T Q M_0(t)\right) dt \\ &+ \int_0^h \left(C^T \bar{f}(t)^T R g(t) C + C^T g(t)^T R \bar{f}(t) C\right) dt \\ P_2 &:= \int_0^h \left(M_1(t)^T Q M_1(t) + C^T g(t)^T R g(t) C\right) dt \end{aligned}$$

Therefore:

$$J = x^T(0) K(\alpha) x(0) \tag{4.20}$$

where:

$$\left(M_0(h)^T + \alpha M_1(h)^T\right) K(\alpha) \left(M_0(h) + \alpha M_1(h)\right) - K(\alpha) + \left(P_0 + P_1\alpha + P_2\alpha^2\right) = 0 \tag{4.21}$$

Note that the equations (4.20) and (4.21) are obtained from (4.14) and (4.15). The problem is now reduced to solving the discrete Lyapunov equation given in (4.21). The following theorem will be used to simplify the problem formulation.

Theorem 1 *The matrix $K(\alpha)$ satisfying (4.21) can be written in the following closed form:*

$$K(\alpha) = \frac{T_0 + T_1\alpha + T_2\alpha^2 + \cdots + T_{2n^2}\alpha^{2n^2}}{r_0 + r_1\alpha + r_2\alpha^2 + \cdots + r_{2n^2}\alpha^{2n^2}} \tag{4.22}$$

where $T_i \in \Re^{n\times n}$, $i = 0,1,...,2n^2$, are constant matrices, and r_i, $i = 0,1,...,2n^2$, are real constant numbers.

Proof One of the approaches for solving a discrete Lyapunov equation is the expansion method, where the matrix equation is expanded into a set of linear algebraic equations. The conventional techniques are then used to solve the resultant equations. This approach is now used to prove Theorem 1. Denote the (i,j) entry of $K(\alpha)$ as $k_{ij}(\alpha)$, for any $i,j \in \{1,2,...,n\}$. Expanding the equation (4.21), n^2 linear equations with n^2 variables $k_{ij}(\alpha)$, $i,j \in \{1,2,...,n\}$, will be obtained. It can be easily verified that each of these linear equations has the following structure:

$$\sum_{1\le i,j\le n} z_v^{ij}(\alpha)k_{ij}(\alpha) = z_v(\alpha), \quad v = 1,2,...,n^2 \tag{4.23}$$

where $z_v^{ij}(\alpha)$, $i,j \in \{1,2,...,n\}$, and $z_v(\alpha)$, $v = 1,2,...,n^2$, are polynomials with degrees of less than or equal to 2. Using Kramer's rule to solve this set of linear equations, one can express $k_{ij}(\alpha)$, $i,j \in \{1,2,...,n\}$, as a fraction whose numerator and denominator orders do not exceed $2n^2$. Furthermore, all of these fractions have the same denominator (because it is, in fact, the determinant of the coefficients matrix). After substituting these expressions into the matrix $K(\alpha)$, the equation (4.22) will be obtained. ■

Theorem 2 *Suppose that r_i, $i = 0,1,2,...,2n^2$, in (4.22) are known. The unknown matrices T_i, $i = 0,1,2,...,2n^2$, can be computed recursively, such that in each recursion one numerical discrete Lyapunov equation is solved.*

Proof Substituting (4.22) into (4.21), results in:

$$\begin{aligned}&\left(M_0(h)^T + \alpha M_1(h)^T\right)\left(T_0 + T_1\alpha + \cdots + T_{2n^2}\alpha^{2n^2}\right)\left(M_0(h) + \alpha M_1(h)\right)\\ &-\left(T_0 + T_1\alpha + \cdots + T_{2n^2}\alpha^{2n^2}\right) + \left(r_0 + r_1\alpha + \cdots + r_{2n^2}\alpha^{2n^2}\right)\left(P_0 + P_1\alpha + P_2\alpha^2\right) = 0\end{aligned} \tag{4.24}$$

One can rearrange the relation given by (4.24) to obtain a polynomial of degree less than or equal to $2n^2 + 2$ with respect to α, where its coefficients are matrices. Since this polynomial is equal to zero for any value of α, all of its coefficients should be zero matrices. This implies that:

$$M_0(h)^T T_0 M_0(h) - T_0 + P_0 r_0 = 0 \tag{4.25a}$$

$$M_0(h)^T T_0 M_1(h) + M_1(h)^T T_0 M_0(h) + M_0(h)^T T_1 M_0(h) - T_1 + P_1 r_0 + P_0 r_1 = 0 \tag{4.25b}$$

$$M_1(h)^T T_{i-2} M_1(h) + M_0(h)^T T_{i-1} M_1(h) + M_1(h)^T T_{i-1} M_0(h) + M_0(h)^T T_i M_0(h) - T_i$$
$$+ P_2 r_{i-2} + P_1 r_{i-1} + P_0 r_i = 0, \quad i = 2,3,...,2n^2 \tag{4.25c}$$

Consider now the equation (4.25a). Since the matrices P_0 and $M_0(h)$ have already been computed and also it is assumed that r_0 is known, this equation is a discrete Lyapunov equation with numeric coefficients, which can be easily solved for T_0. Substituting the computed matrix T_0 into (4.25b) gives another discrete Lyapunov equation, from which T_1 can be obtained. Continuing this procedure, all of the matrices T_i, $i = 0,1,...,2n^2$, will be obtained. As discussed earlier, the discrete closed-loop system is stable for $\alpha = 0$. Thus, all of the eigenvalues of $M_0(h)$ are located inside the unit circle, and consequently, each of the foregoing discrete Lyapunov equations has a unique solution. ■

Theorem 2 gives a recursive method for computing the matrices T_i, $i = 0,1,...,2n^2$, provided the scalars r_i, $i = 0,1,...,2n^2$, are available. It is desired now to present a method for computing these scalars.

Definition 1 *Consider a row vector* $V_1 \in \Re^{n_1}$, *and a column vector* $V_2 \in \Re^{n_2}$. *Define the global multiplication of these two vectors as follows:*

$$GM(V_1,V_2) := \begin{bmatrix} V_1^1.V_2^T & V_1^2.V_2^T & \cdots & V_1^{n_1}.V_2^T \end{bmatrix}^T$$

where V_1^i *represents the* i^{th} *entry of* V_1, $i = 1,2,...,n_1$. *As an example, the global multiplication of* $[1\ 2\ 3]$ *and* $[0\ 1\ 2]^T$ *is* $[0\ 1\ 2\ 0\ 2\ 4\ 0\ 3\ 6]^T$.

Theorem 3 *The scalars* r_i, $i = 0,1,...,2n^2$, *in the denominator of the expression for* $K(\alpha)$ *in (4.22), can be found by computing the matrix* $M_0(h) + \alpha M_1(h)$ *at* $2n^2+1$ *arbitrary and distinct values of* α.

Proof Let the denominator of the expression given in (4.22) be denoted by $\mathrm{den}(\alpha)$. Using (4.23) and (4.21), one can easily verify that:

$$\mathrm{den}(\alpha) = r_0 + r_1\alpha + r_2\alpha^2 + \cdots + r_{2n^2}\alpha^{2n^2}$$
$$= \det\Big(\big[GM\left(E_1(\alpha),F_1(\alpha)\right)\ ,\ ...\ ,\ GM\left(E_1(\alpha),F_n(\alpha)\right)\ ,\ GM\left(E_2(\alpha),F_1(\alpha)\right)\ ,\ ...\ , \tag{4.26}$$
$$GM\left(E_2(\alpha),F_n(\alpha)\right)\ ,\ ...\ ,\ GM\left(E_n(\alpha),F_1(\alpha)\right)\ ,\ ...\ ,\ GM\left(E_n(\alpha),F_n(\alpha)\right)\big] - I_{n^2\times n^2}\Big)$$

(it is to be noted that $\mathrm{den}(\alpha)$ is, in fact, the determinant of the coefficients matrix for the set of linear equations given by (4.23)) where $E_i(\alpha)$ and $F_i(\alpha)$, $i = 1, 2, ..., n$, are the ith row and the ith column of the matrix $M_0(h) + \alpha M_1(h)$, respectively. Consider now $2n^2 + 1$ arbitrary and distinct values for α denoted by $\bar{\alpha}_0, \bar{\alpha}_1, ..., \bar{\alpha}_{2n^2}$. Compute $M_0(h) + \alpha M_1(h)$ for each of these values in order to obtain $E_i(\alpha)$ and $F_i(\alpha)$, and then $\mathrm{den}(\alpha)$ from (4.26). Eventually, the unknown values r_i, $i = 0, 1, ..., 2n^2$, will be obtained as follows:

$$\begin{bmatrix} r_0 \\ r_1 \\ \vdots \\ r_{2n^2} \end{bmatrix} = \begin{bmatrix} 1 & \bar{\alpha}_0 & \bar{\alpha}_0^2 & \dots & \bar{\alpha}_0^{2n^2} \\ 1 & \bar{\alpha}_1 & \bar{\alpha}_2^1 & \dots & \bar{\alpha}_1^{2n^2} \\ \vdots & \vdots & \vdots & \ddots & \vdots \\ 1 & \bar{\alpha}_{2n^2} & \bar{\alpha}_{2n^2}^2 & \dots & \bar{\alpha}_{2n^2}^{2n^2} \end{bmatrix}^{-1} \begin{bmatrix} \mathrm{den}(\bar{\alpha}_0) \\ \mathrm{den}(\bar{\alpha}_1) \\ \vdots \\ \mathrm{den}(\bar{\alpha}_{2n^2}) \end{bmatrix} \tag{4.27}$$

It is to be noted that since $\bar{\alpha}_0, \bar{\alpha}_1, ..., \bar{\alpha}_{2n^2}$ correspond to $2n^2 + 1$ distinct values of α, the matrix given in (4.27) is invertible according to the Vandermond's formula. ∎

As a result, in order to compute $K(\alpha)$, one should first find r_i, $i = 0, 1, ..., 2n^2$, by using Theorem 3, and then compute T_i, $i = 0, 1, ..., 2n^2$, according to Theorem 2.

So far, despite the parametric structure of the discrete Lyapunov equation in (4.21), it is analytically solved by using numerical methods. Substituting the result into (4.20) gives the performance index (4.2) in terms of the unknown variable α and the initial state $x(0)$. Note that for any initial state, the performance index obtained is a rational function whose numerator and denominator are polynomials in α. Note also that to obtain this result, it was assumed that the discrete-time equivalent system corresponding to $\mathscr{S}$ is stable under the GSHF (4.17). Thus, a ROC for α which guarantees the stability of the system $\mathscr{S}$ under the GSHF (4.17) should be obtained. This one-dimensional ROC consists of those values of α for which all of the eigenvalues of the matrix $M_0(h) + \alpha M_1(h)$ are located inside the unit circle. Since it is assumed that $\bar{f}(t)$ stabilizes the system $\mathscr{S}$ for $\alpha = 0$, the point $\alpha = 0$ belongs to the ROC. Moreover, since finding the exact ROC can be cumbersome and very complicated, it is desired to obtain a subset of the exact ROC. This problem is addressed in the following theorem.

Theorem 4 *Let the eigenvalues of $M_0(h)$ be denoted by λ_i, $i = 1, 2, ..., n$, where:*

$$|\lambda_1| \le |\lambda_2| \le \cdots \le |\lambda_n| < 1$$

a) For any $\alpha \in (-\gamma, \gamma)$, where γ is the smallest positive real root of the following polynomial:

$$H(\alpha) = -\frac{(1-|\lambda_n|)^n}{2^{2n-1}} + \alpha\|M_0(h)\|_2\,(2\|M_0(h)\|_2 + \alpha\|M_1(h)\|_2)^{n-1}$$

and where $\|\cdot\|_2$ represents the 2-norm of the corresponding matrix, all of the eigenvalues of the matrix $M_0(h) + \alpha M_1(h)$ are located inside the unit circle.

b) If the matrix $M_0(h)$ is diagonalizable, then γ can be obtained from the following equation:

$$\gamma = \frac{1-|\lambda_n|}{cond_2(V)\|M_1(h)\|_2}$$

where $cond_2(V) := \|V\|_2\|V^{-1}\|_2$ is the condition number of an eigenvector matrix V of $M_0(h)$.

Proof Let the eigenvalues of $M_0(h) + \alpha M_1(h)$ be denoted by λ_i^α, $i = 1, 2, ..., n$. The Elsner theorem [16] states that for any eigenvalue λ_i^α of $M_0(h) + \alpha M_1(h)$, there is an eigenvalue λ_j of $M_0(h)$, such that:

$$|\lambda_i^\alpha - \lambda_j| \le 2^{2-\frac{1}{n}}\|\alpha M_1(h)\|_2^{\frac{1}{n}}\left(\|M_0(h)\|_2 + \|M_0(h) + \alpha M_1(h)\|_2\right)^{1-\frac{1}{n}}$$

Consequently, if α satisfies the following inequality:

$$2^{2-\frac{1}{n}}\|\alpha M_1(h)\|_2^{\frac{1}{n}}\left(2\|M_0(h)\|_2 + \|\alpha M_1(h)\|_2\right)^{1-\frac{1}{n}} < 1 - |\lambda_n|$$

then all of the eigenvalues of the matrix $M_0(h) + \alpha M_1(h)$ will be inside the unit circle. The proof of part (a) follows directly from (4.28), and by noting that this inequality holds for $\alpha = 0$, and does not hold for $\alpha = +\infty$.

On the other hand, if $M_0(h)$ is diagonalizable, the Bauer-Fike theorem [17] states that for any eigenvalue λ_i^α of $M_0(h) + \alpha M_1(h)$, there is an eigenvalue of $M_0(h)$ denoted by λ_j, such that:

$$|\lambda_i^\alpha - \lambda_j| \le \mathrm{cond}_2(V)\|\alpha M_1(h)\|_2 \qquad (4.28)$$

The proof of part (b) follows immediately from (4.28). ■

The bound for α given in part (b) of Theorem 4 is less conservative than the one given in part (a). Hence, when the matrix $M_0(h)$ is diagonalizable, it is better to use the bound given in part (b). It is to be noted that the interval obtained for α in part (b) of Theorem 4 is typically large, because $M_1(h)$ is the integral over the sampling period $[0, h]$, which is typically small, and consequently its 2-norm is small too. It is to be noted that the bounds given in Theorem 4 are symmetric around the origin, due to the nature of the theorems utilized for attaining the bounds. This may lead to a conservative solution, in general. However, one can use the LMI approach proposed in [18; 19] to obtain the exact bound at the cost of more computational effort.

After finding $K(\alpha)$ in terms of α by using Theorems 2 and 3, and substituting them into the equation (4.20), a rational performance index will be obtained with respect to α. Let this performance index be denoted by $J = \frac{\Phi(\alpha)}{\Gamma(\alpha)}$, where Φ and Γ are polynomials of order less than or equal to $2n^2$. The last step of optimization is to minimize this rational function over the interval $(-\gamma, \gamma)$. This problem is often referred to as "global optimization of a constrained rational function" which is, in general, difficult to solve. However, for univariate rational functions, it can be reformulated as a semidefinite programming (SDP) problem [20], which can be solved by several available softwares. It is to be noted that one can directly take the derivative of J with respect to α, equate it to zero, and then find its roots. However, for large values of n, this rudimentary technique is not efficient. Therefore, reformulation of the problem as a SDP will be discussed next. The following lemma is borrowed from [21].

Lemma 5 *Consider the rational function $w(t) = \frac{z(t)}{v(t)}$ over the interval (a,b), and suppose that $z(t)$ and $v(t)$ are relatively prime polynomials. If $v(t)$ changes its sign over (a,b), then the minimum value of the function $w(t)$ over this interval is $-\infty$.*

Consider now the performance index $J = \frac{\Phi(\alpha)}{\Gamma(\alpha)}$ over the interval $(-\gamma, \gamma)$. There exist many simple algorithms to verify whether or not $\Phi(\alpha)$ and $\Gamma(\alpha)$ are relatively prime, and in the case they are not relatively prime, cancel out their greatest common divisor (GCD). Hence, without loss of generality, assume that $\Phi(\alpha)$ and $\Gamma(\alpha)$ are relatively prime polynomials. On the other hand, since the closed-loop system is stable for any $\alpha \in (-\gamma, \gamma)$, the function J should be finite and positive in this interval. Consequently, according to Lemma 5, the sign of the function $\Gamma(\alpha)$ does not change in the interval $(-\gamma, \gamma)$, i.e., it is either always positive or always negative. Now, compute $\Gamma(\alpha)$ for an arbitrary value of α which belongs to the interval $(-\gamma, \gamma)$. If the result is negative, change the signs of the coefficients of both $\Phi(\alpha)$ and $\Gamma(\alpha)$. This conversion does not affect J, but makes the sign of $\Gamma(\alpha)$ positive for all α in $(-\gamma, \gamma)$. As a result, without loss of generality, assume that $\Gamma(\alpha) \geq 0$ for any $\alpha \in (-\gamma, \gamma)$. Let the minimum value of J in this interval be denoted by J_{opt}. One can write:

$$J_{opt} = \sup\{\lambda : \Phi(\alpha) - \lambda\Gamma(\alpha) \geq 0, \ \forall \alpha \in (-\gamma, \gamma)\} \tag{4.29}$$

The above technique is the key to solve the problem of "global optimization of a constrained rational function", because it reduces the optimization of a rational function to the optimization of a polynomial [21].

It follows from the results of M. Fekete theorem [21] and the discussion in Section 3.1.1 of [22], that there exist two positive semidefinite matrices Ω_1 and Ω_2, such that:

$$\Phi(\alpha) - \lambda\Gamma(\alpha) = \tilde{X}_1^T \Omega_1 \tilde{X}_1 + (\gamma^2 - \alpha^2)\tilde{X}_2^T \Omega_2 \tilde{X}_2 \tag{4.30}$$

where

$$\tilde{X}_1 = \begin{bmatrix} 1 & \alpha & \alpha^2 & \cdots & \alpha^{2n^2} \end{bmatrix}^T, \quad \tilde{X}_2 = \begin{bmatrix} 1 & \alpha & \alpha^2 & \cdots & \alpha^{2n^2-1} \end{bmatrix}^T$$

Let the polynomial $\Phi(\alpha) - \lambda\Gamma(\alpha)$ be denoted by $\sum_{i=0}^{2n^2} \zeta_i(\lambda)\alpha^i$. In addition, define Ω_1^{ij} and Ω_2^{ij} as the (i, j) entries of the matrices Ω_1 and Ω_2, respectively. Equating the corresponding coefficients in the two sides of the equation (4.30) yields the following relation for $\nu = 0, 1, ..., 2n^2$:

$$\zeta_\nu(\lambda) = \sum_{i+j=\nu} \Omega_1^{ij} + \gamma^2 \sum_{i+j=\nu} \Omega_2^{ij} - \sum_{i+j=\nu-2} \Omega_2^{ij} \tag{4.31}$$

The optimization problem (4.29) is now equivalent to finding the supremum of λ subject to (4.31), where Ω_1 and Ω_2 are positive semidefinite matrix variables. This is a SDP problem, which can be solved numerically [21], [23].

Algorithm 1

Step 1) *Set* $\alpha_i^0 = \theta_i$, $i = 1, 2, ..., k$, *and* $j = l = 1$.

Step 2) *Set* $g(t) = f_j(t)$.

Step 3) *Compute* $K(\alpha)$ *by using Theorems 3, 2 and 1, and substitute them into (4.20) in order to obtain* J.

Step 4) *Find a region of convergence, denoted by* $(-\gamma, \gamma)$, *using Theorem 4.*

Step 5) *Cancel out the GCD of the numerator and the denominator of J, make the sign of its denominator positive in the interval* $(-\gamma, \gamma)$ *as discussed ealier, and denote the resultant function with* $\frac{\Phi(\alpha)}{\Gamma(\alpha)}$.

Step 6) *Denote* $\Phi(\alpha) - \lambda\Gamma(\alpha)$ *with* $\sum_{i=0}^{2n^2} \zeta_i(\lambda)\alpha^i$. *Now, solve the following optimization problem, which is, in fact, a SDP problem:*

Find the value of α which results in the supremum of λ for the variables Ω_1, Ω_2 and α subject to (4.31), where Ω_1 and Ω_2 are positive semidefinite variables.

Step 7) *Denote the value obtained for* α *in Step 6 with* α_{opt}. *Set* $\alpha_j^l = \alpha_j^{l-1} + \alpha_{opt}$, *and let the new function* $f(t)$ *be the old* $f(t)$ *plus* $\alpha_{opt}g(t)$. *If* $j < k$, *increase* j *by one and go to Step 2.*

Step 8) *If* $\sum_{i=1}^{k} |\alpha_i^l - \alpha_i^{l-1}| > \delta$, *where* δ *is a prescribed tolerance, then set* $j = 1$, *increase the value of* l *by one, and go to Step 2.*

Step 9) *The optimal coefficients of the GSHF* $f(t)$ *defined in (4.5) are given by* $\alpha_i = \alpha_i^l$, $i = 1, 2, ..., k$.

It is to be noted that Algorithm 1 should be ideally halted when $\alpha_i^l = \alpha_i^{l-1}$, $i = 1, 2, ..., k$. However, Step 8 is added to the algorithm to assure that it will stop in a finite time.

Remark 2 *The problem of finding the optimal value of* α *is, in fact, formulated as a constrained optimization problem by concentrating on only the values of* α *in the open interval* $(-\gamma, \gamma)$. *However, if the solution of this problem is* $-\gamma^+$ *or* γ^-, *this implies that the global minimum of the performance index most likely corresponds to a value of* α *outside the region given by Theorem 4. In other words, if the minimum value of the above constrained optimization problem occurs at* $\alpha = -\gamma^+$ *or* $\alpha = \gamma^-$, *the global minimum of the unconstrained optimization problem (where the only implicit constraint is stability), occurs for a value of* α *less than or equal to* γ, *or greater than or equal equal to* $-\gamma$, *respectively. To take these case into consideration, one can replace Step 7 of Algorithm 1 with the following:*

Step 7) Denote the value obtained for α in Step 6 with α_{opt}. Set $\alpha_i^l = \alpha_i^{l-1} + \alpha_{opt}$, and the new function $f(t)$ as the old $f(t)$ plus $\alpha_{opt}g(t)$. If $\alpha_{opt} = -\gamma^+$ or $\alpha_{opt} = \gamma^-$, go to Step 2. If $i < k$, increase the value of i by one and go to Step 2.

Remark 3 *As discussed earlier, the optimal value of α depends on the initial state $x(0)$. However, if the exact initial state is unknown or if it is desired to find a value of α which is independent of the initial state, the optimization problem can be treated probabilistically. Assume that the expected value of $x(0)x(0)^T$ is known and denoted by X_0. Suppose that it is desired to minimize the expected value of J over all initial values. One can write:*

$$E\{J\} = E\left\{x^T(0)K(\alpha)x(0)\right\} = trace\Big(K(\alpha)X_0\Big) \tag{4.32}$$

If $K(\alpha)$ (which is obtained by using the method discussed earlier) is substituted into the above equation, a new rational function will be obtained. Now, the probabilistic optimization problem reduces to minimizing this rational function over the interval $(-\gamma,\gamma)$, which can be treated by using the SDP approach as discussed before.

Remark 4 *After formulating the problem with the equations (4.20) and (4.21), one can choose any converging numerical algorithm to find the optimal value of α. However, the problem is not convex, and also the step sizes required for convergence to a point with a desired accuracy is unknown. Hence, one of the novelties of the approach presented in this chapter is that it specifies the step sizes (except for Step 6 of Algorithm 1 regarding the optimization of a rational function that generally converges to its optimal point very fast compared to the algorithm presented in [13; 10]).*

4.5 Numerical examples

Example 1 (Harmonic Oscillator) Consider a continuous-time LTI system with the following state-space matrices:

$$A = \begin{bmatrix} 0 & \zeta \\ -\zeta & 0 \end{bmatrix}, \quad B = \begin{bmatrix} 0 \\ 1 \end{bmatrix}, \quad C = \begin{bmatrix} 1 & 0 \end{bmatrix} \tag{4.33}$$

and $x(0) = [\ 1 \quad 1\ ]^T$. Two different values $(1,2)$ and $(-5,1)$ for the pair (ζ,h) will be considered here, which correspond to the examples presented in [13] and [12], respectively.

i) $\zeta = 1$, $h = 2$sec: If a ZOH is used with a unity controller, it can be easily verified that the overall closed-loop system will be unstable. Note that for ZOH, $f(t)$ is equal to 1. Suppose that it is desired to obtain a constant GSHF, which minimizes the performance index (4.2), with $Q = R = I$. To find the desired GSHF $\bar{f}(t)$, an initial constant stabilizing GSHF $f(t)$ is required. One can verify that

$f(t) = 0.9$ stabilizes the system. For this constant GSHF, the performance index is 28.199. Let the function $g(t)$ given in (4.17) be equal to 1. Pursuing the proposed method, the optimal GSHF $\bar{f}(t)$ is obtained to be 0.502, which yields $J = 9.078$. This means that the minimum achievable performance index is 9.078 under the constraint of constant GSHF. It is to be noted that a constant GSHF is equivalent to a ZOH and a constant gain controller.

Suppose now that it is desired to use a piecewise constant GSHF to further improve the performance index. Consider the following three basis functions for the desired GSHF:

$$f_i(t) = u_1\left(t - \frac{(i-1)h}{3}\right) - u_1\left(t - \frac{ih}{3}\right), \quad i = 1,2,3 \tag{4.34}$$

where $u_1(\cdot)$ represents the unit-step function. One can commence the proposed procedure from the values $\theta_1 = \theta_2 = \theta_3 = 0.9$ in (4.5). Tuning these coefficients by using Algorithm 1 yields $\alpha_1 = -0.399$, $\alpha_2 = 0.690$ and $\alpha_3 = 0.903$. Hence, the optimal piecewise constant GSHF with the basis functions (4.34) is as follows:

$$\bar{f}(t) = -0.399 f_1(t) + 0.690 f_2(t) + 0.903 f_3(t) \tag{4.35}$$

and the corresponding optimal performance index is 7.088.

The unconstrained optimal GSHF for this system is derived in [13]. However, since it is obtained through an iterative procedure, the optimal GSHF is a curve for which no judgement can be made (because the function does not have any closed-form expression). However, the performance index corresponding to the GSHF given in [13] is 6.8. In contrast, by using a very simple structure for GSHF as discussed above, the resultant performance index (7.088) is very close to the unconstrained optimal value (6.8).

ii) $\zeta = -5,\ h = 1\text{sec}$: Suppose that it is desired to have a piecewise constant GSHF with the basis functions:

$$f_i(t) = u_1\left(t - \frac{(i-1)h}{2}\right) - u_1\left(t - \frac{ih}{2}\right), \quad i = 1,2 \tag{4.36}$$

Starting from $\theta_1 = \theta_2 = -4$, the optimal GSHF is obtained to be:

$$\bar{f}(t) = -1.371 f_1(t) + 0.960 f_2(t) \tag{4.37}$$

and the corresponding optimal performance index will be 3.474. On the other hand, using the optimal continuous-time LQR controller and assuming that both states are available in the output (i.e., without using an observer), the minimum performance index will be 2.651 (note that this is, in fact, the minimum achievable performance index for the system using any type of control). The output of the system under the optimal continuous-time LQR controller and under the optimal GSHF (4.37) are depicted in Figure 4.1. This figure demonstrates that the proposed GSHF performs very closely to the optimal LQR controller. This is due to the fact that by using GSHF, one can obtain much of the efficiency of state feedback, without the requirement of state estimation [3].

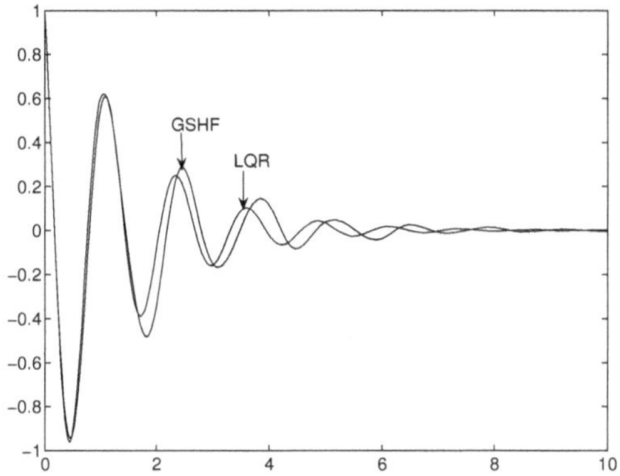

Figure 4.1: The output of the system in Example 1 under the continuous-time optimal LQR controller and under the optimal GSHF given in (4.37).

Example 2 Consider a continuous-time LTI system with the following state-space matrices:

$$A = \begin{bmatrix} -1 & -4 \\ 5 & 3 \end{bmatrix}, \quad B = \begin{bmatrix} -3 & 0 \\ 0 & 1 \end{bmatrix}, \quad C = \begin{bmatrix} 1 & 0 \\ 0 & 1 \end{bmatrix}$$

and the initial state $x(0) = [\ 1 \quad 1\ ]^T$. Let the sampling period be $h = 0.1$sec. It can be easily verified that the minimum achievable performance index (4.2) with $Q = R = I$ for the system is 2.857. Assume now that it is desired to design a decentralized periodic controller for the system. In other words, it is required to obtain a 2×2 diagonal GSHF. Furthermore, assume that the basis functions for the GSHF are to be:

$$f_1(t) = \begin{bmatrix} 1 & 0 \\ 0 & 0 \end{bmatrix}, \quad f_2(t) = \begin{bmatrix} 0 & 0 \\ 0 & 1 \end{bmatrix}, \quad f_3(t) = \begin{bmatrix} \sin(t) & 0 \\ 0 & 0 \end{bmatrix}$$

Let the algorithm start with the initial parameters $\theta_1 = 1,\ \theta_2 = 1,\ \theta_3 = 0$. Note that these parameters represent a simple ZOH for each input-output agent in the closed-loop configuration, and result in an initial performance index equal to 475.110. Now, let the coefficients of these three basis functions be tuned as proposed in the present chapter. This will result in:

$$\bar{f}(t) = \begin{bmatrix} 1.084 + 27.128\sin(t) & 0 \\ 0 & -1.360 \end{bmatrix} \tag{4.38}$$

Consequently, the resultant performance index will be 5.396. This implies that the decentralized performance index is improved from 475.110 to 5.396, while the minimum achievable performance index (with no structural constraint) is 2.8566.

4.6 Conclusions and suggested future work

In this chapter, a novel approach is proposed to design a generalized sampled-data hold function (GSHF) with any prespecifed structure (e.g., decentralized with block diagonal information flow matrix) and any given form (e.g., polynomial) for a continuous-time finite-dimensional linear time-invariant system. The resultant GSHF is optimal with respect to a linear-quadratic cost function, subject to the constraints imposed on the structure and the form of the GSHF. A necessary and sufficient condition for the existence of a stailizing GSHF with the desired constraints is obtained. Then, an algorithm is proposed to find the optimal GSHF. This chapter uses the recent results in semidefinite programming. Simulation results demonstrate the effectiveness of the proposed method.

There are several suggestions as the continuation of the work proposed in present chapter. First, it would be very useful to find an optimal choice of the set of basis functions with any given size (in terms of the number of functions in the set). Another direction for the future work, could be a systematic methodology to obtain a set of basis functions which can potentially stabilize a large-scale system by means of a decentralized controller, when there is no LTI decentralized controller to achieve stability. Furthermore, using H_∞ norm instead of H_2 as the performance index can be interesting as far as robustness is concerned.

Bibliography

[1] J. I. Yuz, G. C. Goodwin and H. Garnier, "Generalised hold functions for fast sampling rates," *in Proc. 43rd IEEE Conf. on Decision and Contr.*, Atlantis, Bahamas, vol. 2, pp. 1908-1913, 2004.

[2] M. Rossi and D. E. Miller, "Gain/phase margin improvement using static generalized sampled-data hold functions," *Sys. and Contr. Lett.*, vol. 37, no. 3, pp. 163-172, 1999.

[3] P. T. Kabamba, "Control of linear systems using generalized sampled-data hold functions," *IEEE Trans. on Automat. Contrl.*, vol. 32, no. 9, pp. 772-783, 1987.

[4] A. Feuer and G. C. Goodwin, "Generalized sample hold functions-frequency domain analysis of robustness, sensitivity, and intersample difficulties," *IEEE Trans. on Automat. Contrl.*, vol. 39, no. 5, pp. 1042-1047, 1994.

[5] K. G. Arvanitis, "On the localization of intersample ripples of linear systems controlled by generalized sampled-data hold functions," *Automatica*, vol. 34, no. 8, pp. 1021-1024, 1998.

[6] E. J. Davison and T. N. Chang, "Decentralized stabilization and pole assignment for general proper systems," *IEEE Trans. on Automat. Contrl.*, vol. 35, no. 6, pp. 652-664, June 1990.

[7] J. L. Willems, "Time-varying feedback for the stabilization of fixed modes in decentralized control systems," *Automatica*, vol. 25, no. 1, pp. 127-131, 1989.

[8] Z. Gong and M. Aldeen, "Stabilization of decentralized control systems," *Journal of Mathematical Systems, Estimation, and Control*, vol. 7, no. 1, pp. 1-16, 1997.

[9] S. H. Wang, "Stabilization of decentralized control systems via time-varying controllers," *IEEE Trans. on Automat. Contrl.*, vol. 27, no. 3, pp. 741-744, Jun. 1982.

[10] Y. Y. Cao and J. Lam, "A computational method for simultaneous LQ optimal control design via piecewise constant output feedback," *IEEE Trans. on Sys. Man Cyber.*, vol. 31, no. 5, pp. 836-842, Oct. 2001.

[11] A. G. Aghdam, "Decentralized control design using piecewise constant hold functions," *in Proc. 2006 American Contr. Conf.*, Minneapolis, Minnesota, 2006.

[12] Y. C. Juan and P.T. Kabamba, "Optimal hold functions for sampled data regulation," *Automatica*, vol. 27, no. 1, pp. 177-181, 1991.

[13] G. L. Hyslop, H. Schattler and T. J. Tarn, "Descent algorithms for optimal periodic output feedback control," *IEEE Trans. on Automat. Contrl.*, vol. 37, no. 12, pp. 1893-1904, 1992.

[14] H. Werner, "An iterative algorithm for suboptimal periodic output feedback control," *UKACC Int. Conf. Contrl.*, vol. 2, pp. 814-818, 1996.

[15] A. Isaacs, Direct-search methods and DACE, Ph.D. Thesis, Indian Institute of Technology Bombay, 2003.

[16] R. Bhatia, L. Elsner and G. Krause, "Bounds for variation of the roots of a polynomial and the eigenvalues of a matrix," *Lin. Algeb. Appl.*, vol. 142, pp. 195-209, 1990.

[17] S. C. Eisenstat and C. Ipsen, "Three absolute perturbation bounds for matrix eigenvalues imply relative bounds," *SIAM J. Matr. Anal. & Appl.*, vol. 20, no. 1, pp. 149-159, 1998. 6, pp. 652-664, June 1990.

[18] J. Lavaei and A. G. Aghdam, "Robust stability of LTI discrete-time systems using sum-of-squares matrix polynomials," *in Proc. 2006 American Contr. Conf.*, pp. 3828-3830, Minneapolis, MN, 2006.

[19] J. Lavaei and A. G. Aghdam, "A necessary and sufficient condition for robust stability of LTI discrete-time systems using sum-of-squares matrix polynomials," *in Proc. of 45th IEEE Conf. on Decision and Contr.*, San Diego, CA, 2006.

[20] L. Vandenberghe and S. Boyd, "Semidefinite programming," *SIAM Rev.*, vol. 38, no. 1, pp. 49-95, 1996.

[21] D. Jibetean and E. de Klerk, "Global optimization of rational functions: a semidefinite programming approach," *Mathematical Programming*, vol. 106, no. 1, pp. 93-109, May 2006.

[22] D. Jibetean. Algebraic optimization with applications to system theory, Ph.D. Thesis, Vrije Universiteit, 2003.

[23] D. Henrion and J. B. Lasserre, "GloptiPoly: global optimization over polynomials with Matlab and SeDuMi," *ACM Trans. on Math. Soft.*, vol. 29, no. 2, pp. 165-194, 2003.

Chapter 5

Simultaneous Stabilization using Decentralized Periodic Feedback Law

5.1 Abstract

In this chapter, controlling a set of continuous-time LTI systems is considered. It is assumed that a predefined guaranteed continuous-time quadratic cost function, which is, in fact, the summation of the performance indices for all systems, is given. The main objective here is to design a decentralized periodic output feedback controller with a prespecified form, e.g., polynomial, piecewise constant, exponential, etc., which minimizes the above mentioned guaranteed cost function. This problem is first formulated as a set of matrix inequalities, and then by using a well-known technique, it is reformulated as a LMI problem. The set of linear matrix inequalities obtained represent the necessary and sufficient conditions for existence of the desired structurally constrained controller. Moreover, an algorithm is presented to solve the resultant LMI problem. Finally, the efficiency of the proposed method is demonstrated in two numerical examples, which are investigated in several relevant papers.

5.2 Introduction

The idea of using generalized sampled-data hold functions (GSHF) instead of a simple zero-order hold (ZOH) in control systems was first introduced in [1]. Kabamba investigated several applications and properties of GSHF in control systems [2; 3; 4]. He showed that many of the advantages of state feedback controllers, without the requirement of using state estimation procedures can be obtained by using a GSHF. A comprehensive frequency-domain analysis was done by Feuer and Goodwin to examine robustness, sensitivity, and intersample effect of GSHF [5]. Several advantages and disadvantages of GSHF and its application in practical problems have been thoroughly investigated and different design techniques are proposed in the literature [6; 7].

Simultaneous stabilization of a set of systems, on the other hand, is of special interest in the control literature [8; 9], and has applications in the following problems:

- A system which is desired to be stabilized by a fixed controller in different modes of operations, e.g., failure mode.
- A nonlinear plant which is linearized at several equilibria.
- A system which is desired to be stabilized in presence of uncertainties in its parameters.

Despite numerous efforts made to solve the simultaneous stabilization problem, it still remains an open problem. In the special case, when there are only two plants to be simultaneously stabilized, the problem is completely solved in [10; 11], and for the case of three and four plants, some necessary and sufficient relations in the form of polynomial are presented in [12]. However, no necessary and sufficient condition has been obtained for simultaneous stabilization of more than four plants, so far. Moreover, it is proved in [13] that if the number of plants is more than two, then the problem is rationally undecidable. It is also shown in [14] that the problem is NP-hard. These results clearly demonstrate complexity level of the problem. Since there does not exist any LTI simultaneous stabilizing controller in many cases, a time-varying controller is considered in [15]. It is shown that for any set of stabilizable and observable plants, there exists a time-varying controller consisting of a sampler, a ZOH, and a time-varying discrete-time compensator which not only stabilizes all of the plants, but also acts as a near-optimal controller for each plant. This result points to the usefulness of sampling in simultaneous stabilization problem. Nevertheless, fast sampling requirement and large control gain are the drawbacks of this approach.

Stabilizing a set of plants simultaneously by means of a periodic controller in order to achieve good behavior for the control systems is investigated in the literature [4; 16]. A method is proposed in [4] to not only minimize a guaranteed cost function corresponding to all of the systems, but also accomplish desired pole placement. The drawback, however, is that the problem is formulated as a two-boundary point differential equation whose analytical solution is cumbersome, in general. Some algorithms are proposed to solve the resultant differential equation numerically, in the particular case of only one LTI system, which is no longer a simultaneous stabilization problem [17; 18]. Design of a high-performance simultaneous stabilizer in the form of a piecewise constant GSHF is investigated in [19].

On the other hand, it is not realistic in many practical problems to assume that all of the outputs of a system are available to construct any particular input of the system. In other words, it is often desired to have some form of decentralization. Problems of this kind appear, for example, in electric power systems, communication networks, large space structures, robotic systems, economic systems and traffic networks, to name only a few. Note that throughout this chapter the term "decentralized control" refers to a controller which constructs any input of the system using certain outputs, determined by the given information flow structure [20; 21].

This chapter deals with the problem of simultaneous stabilization of a set of systems by means of a decentralized periodic controller. It is assumed that a discrete-time decentralized compensator is given for a set of detectable and stabilizable LTI systems. This compensator is employed to simplify the simultaneous stabilizer design problem. In certain cases, however, the problem may not be solvable without using a proper compensator (e.g., in presence of unstable fixed modes [21]). The objective is to design a GSHF which satisfies the following constraints:

i) The GSHF along with the discrete-time compensator simultaneously stabilize the plants.

ii) It has the desired decentralized structure.

iii) It has a prespecified form such as polynomial, piecewise constant, etc.

iv) It minimizes a predefined guaranteed cost function, which is the summation of the performance indices of all plants.

It is to be noted that condition (iii) given above is motivated by the following practical issues:

- In many problems involving robustness, noise rejection, simplicity of implementation, elimination of fixed modes, etc., it is desired to design GSHFs with a specific form, e.g. piecewise constant, exponential, etc. [2; 22; 23].

- Design of a high performance simultaneously stabilizing *piecewise constant* GSHF with no compensator is studied in [19]. Therefore, the present chapter solves the most general form of the problem.

- In the case of sufficiently small sampling period, the optimal simultaneous stabilizer (whose exact solution, as pointed out earlier, involves complicated computations), can be approximated by a polynomial (e.g., the truncated Taylor series).

Conditions (ii) and (iii) are formulated by writing the GSHF as a linear combination of appropriate basis functions. The problem is then reduced to finding the coefficients of the linear combination, such that conditions (i) and (iv) are met. It is shown that the aforementioned problem is solvable (i.e., the desired GSHF and compensator exist), if and only if a particular set of systems are simultaneously stabilizable by means of a decentralized *static* output feedback, which unlike the general simultaneous stabilization problem, is rationally decidable [14]. Furthermore, one of the substantial features of the present work is that in order to improve the performance of the system, one can simply extend the set of the basis functions and find the corresponding new coefficients, accordingly. On the contrary, in the case of continuous-time LTI controller, the order of the controller needs to be increased (which increases the complexity of the overall system) in order to achieve a higher performance. Moreover, it is shown in an example that there may exist no LTI controller to stabilize the system, while a simple GSHF can stabilize it and result in an excellent performance.

5.3 Problem formulation

Consider a set of η continuous-time detectable and stabilizable LTI systems $\mathscr{S}_1, \mathscr{S}_2, ..., \mathscr{S}_\eta$ with the following state-space representations:

$$\dot{x}_i(t) = A_i x_i(t) + B_i u_i(t) \tag{5.1a}$$

$$y_i(t) = C_i x_i(t) \tag{5.1b}$$

where $x_i \in \Re^{n_i}$, $u_i \in \Re^m$ and $y_i \in \Re^l$, $i \in \bar{\eta} := \{1, 2, ..., \eta\}$, are the state, the input and the output of $\mathscr{S}_i$, respectively. Assume that the discrete-time compensator K_c^i, $i \in \bar{\eta}$, with the following representation is given:

$$z_i[\kappa+1] = Ez_i[\kappa] + Fy_i[\kappa]$$
$$\phi_i[\kappa] = Gz_i[\kappa] + Hy_i[\kappa] \tag{5.2}$$

and assume also that $z_i[0] = 0$. It is to be noted that the discrete argument corresponding to the samples of any signal is enclosed in brackets (e.g., $y_i[\kappa] = y_i(\kappa h)$). K_c^i can be either decentralized with block-diagonal transfer function matrix or centralized. Suppose now that the system $\mathscr{S}_i$, $i \in \bar{\eta}$ is desired to be controlled by the compensator K_c^i and the hold controller K_h^i represented by:

$$u_i(t) = f(t)\phi_i[\kappa], \quad \kappa h \leq t < (\kappa+1)h, \quad \kappa = 0, 1, 2, \ldots$$

where h is the sampling period, and $f(t) = f(t+h)$, $t \geq 0$. Note that $f(t)$ is a sampled-data hold function, which is desired to be described by the following set of basis functions:

$$\mathbf{f} := \{f_1(t), f_2(t), \ldots, f_k(t)\}$$

where $f_i(t) \in \Re^{m \times l_i}$, $i = 1, 2, \ldots, k$. Thus, $f(t)$ can be written as a linear combination of the basis functions in $\mathbf{f}$ as follows:

$$f(t) = f_1(t)\alpha_1 + f_2(t)\alpha_2 + \cdots + f_k(t)\alpha_k \tag{5.3}$$

where some of the entries of the variable matrices $\alpha_i \in \Re^{l_i \times l}$, $i = 1, 2, \ldots, k$, are set equal to zero and the other entries are free variables so that the structure of $f(t)$ complies with the desired control constraint, which is determined by a given information flow matrix [21]. This is illustrated later in Example 2. Furthermore, the set of basis functions $\mathbf{f}$ is obtained according to the desirable form of GSHF (e.g, exponential, polynomial, etc.). This will be demonstrated in Examples 1 and 2. Note that the motivation for considering a special form for $f(t)$ is discussed in the introduction. Besides, some examples are presented in [24] to demonstrate the effectiveness of the proposed formulation for GSHF.

For any $i \in \{1, 2, \ldots, k\}$, put all of the indices of the zeroed entries of α_i in the set $\mathbf{E_i}$. Assume now the expected value of $x_i(0)x_i(0)^T$, which is referred to as the covariance matrix of the initial state $x_i(0)$, is known and denoted by X_0^i for any $i \in \bar{\eta}$. The objective is to obtain the constrained matrices $\alpha_1, \ldots, \alpha_k$ such that the following performance index is minimized:

$$J = E\left\{ \sum_{i=1}^{\eta} \int_0^{\infty} \left(x_i(t)^T Q_i x_i(t) + u_i(t)^T R_i u_i(t)\right) dt \right\} \tag{5.4}$$

where $R_i \in \Re^{m\times m}$ and $Q_i \in \Re^{n_i\times n_i}$ are symmetric positive definite and symmetric positive semi-definite matrices, respectively, and $E\{\cdot\}$ denotes the expectation operator. Note that by minimizing the cost function given above, the stability of the system $\mathscr{S}_i$ under the discrete-time compensator K_c^i and the hold controller K_h^i, for any $i \in \bar{\eta}$, is achieved because the cost function becomes infinity otherwise. Note also that since (5.4) is a continuous-time performance index, it takes the intersample ripple effect into account.

The equation (5.3) can be written as $f(t) = g(t)\alpha$, where:

$$g(t) := [f_1(t)\ f_2(t)\ \dots\ f_k(t)],\ \ \alpha := \begin{bmatrix}\alpha_1^T & \alpha_2^T & \dots & \alpha_k^T\end{bmatrix}^T \tag{5.5}$$

Define a new set $\mathbf{E}$ based on the sets $\mathbf{E_1}, \dots, \mathbf{E_k}$, such that any of the entries of α whose index belongs to $\mathbf{E}$ is equal to zero. On the other hand, it is known that:

$$x_i(t) = e^{(t-\kappa h)A_i}x_i(\kappa h) + \int_{\kappa h}^{t} e^{(t-\tau)A_i}B_iu_i(\tau)d\tau$$

for any $\kappa h \le t \le (\kappa+1)h,\ \kappa \ge 0$. Let the following matrices be defined for any $i \in \bar{\eta}$:

$$M_i(t) := e^{tA_i},\ \ \bar{M}_i(t) := \int_0^t e^{(t-\tau)A_i}B_ig(\tau)d\tau$$

Therefore:

$$x_i(t) = M_i(t-\kappa h)x_i[\kappa] + \bar{M}_i(t-\kappa h)\alpha\phi_i[k] \tag{5.6}$$

for any $\kappa h \le t \le (\kappa+1)h$. It can be easily concluded from (5.1b), (5.2), and (5.6) by substituting $t = (\kappa+1)h$, that $\mathbf{x}_i[\kappa+1] = \tilde{M}_i(h,\alpha)\mathbf{x}_i[\kappa]$ for any $\kappa \ge 0$, where $\mathbf{x}_i[\kappa] = \begin{bmatrix} x_i[\kappa]^T & z_i[\kappa]^T \end{bmatrix}^T$, and:

$$\tilde{M}_i(h,\alpha) := \begin{bmatrix} M_i(h) + \bar{M}_i(h)\alpha HC_i & \bar{M}_i(h)\alpha G \\ FC_i & E \end{bmatrix} \tag{5.7}$$

It is straightforward to show that:

$$\mathbf{x}_i[\kappa] = \left(\tilde{M}_i(h,\alpha)\right)^{\kappa}\mathbf{x}_i[0],\quad \kappa = 0,1,2,\dots$$

5.4 Optimal Structurally Constrained GSHF

It is desired now to find out when the structurally constrained GSHF $f(t)$ exists such that the system $\mathscr{S}_i$ is stable under the compensator K_c^i and the hold controller K_h^i, for any $i \in \bar{\eta}$.

Lemma 1 *There exists a GSHF $f(t)$ with the desired form (given by the equation (5.3)) such that the system $\mathscr{S}_i$ is stable under the compensator K_c^i and the hold controller K_h^i for any $i \in \bar{\eta}$, if and only if there exists an output feedback with the constant gain α, with the properties that:*

1. *Each entry of α whose index belongs to the set $\mathbf{E}$ is equal to zero.*

2. *It simultaneously stabilizes all of the η systems $\bar{\mathscr{S}}_1, \bar{\mathscr{S}}_2, ..., \bar{\mathscr{S}}_\eta$, where the system $\bar{\mathscr{S}}_i$, $i \in \bar{\eta}$, is represented by:*

$$\bar{x}_i[\kappa+1] = \begin{bmatrix} M_i(h) & 0 \\ FC_i & E \end{bmatrix} \bar{x}_i[\kappa] + \begin{bmatrix} \bar{M}_i(h) \\ 0 \end{bmatrix} \bar{u}_i[\kappa]$$

$$\bar{y}_i[\kappa] = \begin{bmatrix} HC_i & G \end{bmatrix} \bar{x}_i[\kappa]$$

(note that each of the two 0's in the above equation represents a zero matrix with proper dimension).

Proof The proof follows from the fact that the system $\mathscr{S}_i$, $i \in \bar{\eta}$, is stable under K_c^i and K_h^i if and only if all of the eigenvalues of the matrix $\tilde{M}_i(h, \alpha)$ given in (5.7) are located inside the unit circle in the complex plane. ∎

Remark 1 *Lemma 1 presents a necessary and sufficient condition for the existence of a structurally constrained GSHF $f(t)$ with a desired form, which simultaneously stabilizes all of the systems $\mathscr{S}_1, \mathscr{S}_2$, ..., $\mathscr{S}_\eta$ along with a given discrete-time compensator. The condition obtained is usually referred to as "simultaneous stabilization of a set of LTI systems via structured static output feedback", which has been investigated intensively in the literature. For instance, one can exploit the LMI algorithm proposed in [19] to solve the simultaneous stabilization problem given in Lemma 1 in order to obtain a stabilizing matrix α denoted by $\check{\alpha}$ (which is later used as the initial point in the main algorithm), or conclude the non-existence of such GSHF, otherwise.*

Define now the following matrices for any $i \in \bar{\eta}$:

$$P_0^i := \int_0^h \left(M_i(t)^T Q_i M_i(t)\right) dt$$
$$P_1^i := \int_0^h \left(M_i(t)^T Q_i \bar{M}_i(t)\right) dt$$
$$P_2^i := \int_0^h \left(\bar{M}_i(t)^T Q_i \bar{M}_i(t) + g(t)^T R_i g(t)\right) dt$$
$$q_0^i(\alpha) := P_0^i + P_1^i \alpha H C_i + (P_1^i \alpha H C_i)^T + (\alpha H C_i)^T P_2^i (\alpha H C_i)$$
$$q_1^i(\alpha) := P_1^i \alpha G + (\alpha H C)^T P_2^i \alpha G$$
$$N_i(\alpha) := \begin{bmatrix} q_0^i(\alpha) & q_1^i(\alpha) \\ q_1^i(\alpha)^T & G^T \alpha^T P_2^i \alpha G \end{bmatrix}$$

Theorem 1 *Suppose that the system $\mathscr{S}_i$ is stable under K_c^i and K_h^i, for any $i \in \bar{\eta}$. The performance index J defined in (5.4) can be written as:*

$$J = trace\left(\sum_{i=1}^{\eta} K_i \begin{bmatrix} X_0^i & 0 \\ 0 & 0 \end{bmatrix}\right) \tag{5.8}$$

where K_i, $i \in \bar{\eta}$, satisfies the following discrete Lyapunov equation:

$$\tilde{M}_i^T(h,\alpha) K_i \tilde{M}_i(h,\alpha) - K_i + N_i(\alpha) = 0 \tag{5.9}$$

Proof One can write the performance index as follows:

$$J = E\left\{\sum_{i=1}^{\eta}\sum_{\kappa=0}^{\infty}\int_{\kappa h}^{(\kappa+1)h}\left(x_i^T(t) Q_i x_i(t) + u_i^T(t) R_i u_i(t)\right) dt\right\}$$
$$= E\left\{\sum_{i=1}^{\eta}\sum_{\kappa=0}^{\infty}\left(\mathbf{x}_i^T[\kappa] N_i(\alpha) \mathbf{x}_i[\kappa]\right)\right\}$$
$$= E\left\{\sum_{i=1}^{\eta}\mathbf{x}_i^T(0) K_i \mathbf{x}_i(0)\right\}$$

where:

$$K_i = \sum_{\kappa=0}^{\infty} \tilde{M}_i^T(h,\alpha)^{\kappa} N_i(\alpha) \tilde{M}_i(h,\alpha)^{\kappa} \tag{5.10}$$

Since it is assumed that the system $\mathscr{S}_i$, $i \in \bar{\eta}$, is stable under K_c^i and K_h^i, it can be concluded from the discussion in the proof of Lemma 1 that all of the eigenvalues of the matrix $\tilde{M}_i(h,\alpha)$, $i \in \bar{\eta}$, are located inside the unit circle in the complex plane. This implies the convergence of the infinite series given in (5.10), where its solution satisfies the discrete Lyapunov equation (5.9). This completes the proof. ■

The following lemma reformulates the problem of minimizing the performance index defined by (5.4) (or equivalently (5.8)).

Lemma 2 *Assume that all of the eigenvalues of the matrix $\tilde{M}_i(h,\alpha)$ lie inside the unit circle.*

a) Consider an arbitrary matrix K_i^*, such that:

$$\tilde{M}_i^T(h,\alpha)K_i^*\tilde{M}_i(h,\alpha) - K_i^* + N_i(\alpha) < 0 \tag{5.11}$$

Then $\mathbf{x}_i(0)^T K_i \mathbf{x}_i(0) \leq \mathbf{x}_i(0)^T K_i^* \mathbf{x}_i(0)$.

b) For any number ζ greater than $\mathbf{x}_i(0)^T K_i \mathbf{x}_i(0)$, there exists a positive definite matrix K_i^* satisfying the inequality (5.11) such that $\mathbf{x}_i(0)^T K_i^* \mathbf{x}_i(0) < \zeta$.

Proof *Proof of (a):* Suppose that the matrix K_i^* satisfies the inequality (5.11). It can be concluded from (5.9) that:

$$\tilde{M}_i^T(h,\alpha)(K_i^* - K_i)\tilde{M}_i(h,\alpha) - (K_i^* - K_i) < 0 \tag{5.12}$$

Since all of the eigenvalues of the matrix $\tilde{M}_i(h,\alpha)$ are inside the unit circle, it can be concluded from the above inequality that $K_i^* - K_i$ is positive semidefinite. Hence:

$$\mathbf{x}_i(0)^T (K_i^* - K_i)\mathbf{x}_i(0) \geq 0 \tag{5.13}$$

or equivalently:

$$\mathbf{x}_i(0)^T K_i \mathbf{x}_i(0) \leq \mathbf{x}_i(0)^T K_i^* \mathbf{x}_i(0) \tag{5.14}$$

Proof of (b): Consider the following discrete Lyapunov equation:

$$\tilde{M}_i^T(h,\alpha)\Omega_\varepsilon\tilde{M}_i(h,\alpha) - \Omega_\varepsilon + \varepsilon I = 0 \tag{5.15}$$

Since all of the eigenvalues of the matrix $\tilde{M}_i(h,\alpha)$ are located inside the unit circle, for any $\varepsilon > 0$, there exists a unique matrix $\Omega_\varepsilon > 0$ that satisfies the above equation. Define now the matrix $K_i^* := \Omega_\varepsilon + K_i$. It is clear that K_i^* satisfies (5.12) and also (5.11). The upper bound for the solution of a discrete Lyapunov equation presented in [25] yields the inequality $\|\Omega_\varepsilon\| \leq \varepsilon \times s(h,\alpha)$, where $\|\cdot\|$ represents the Frobenius norm, and the function $s(h,\alpha)$ is related to $\tilde{M}_i(h,\alpha)$. Therefore, it can be concluded that as ε goes to zero, the matrix Ω_ε approaches the zero matrix. As a result, when ε goes to zero, the matrix K_i^* converges to the matrix K_i. This means that the term $\mathbf{x}_i(0)^T K_i^* \mathbf{x}_i(0)$ can be made sufficiently close to $\mathbf{x}_i(0)^T K_i \mathbf{x}_i(0)$. ■

According to Lemma 2, the problem of minimizing J given by (5.8) subject to the constraint (5.9) can be equivalently considered as the minimization of J subject to $\tilde{M}_i^T(h,\alpha)K_i\tilde{M}(h,\alpha)-K_i+N_i(\alpha)<0$, for $i=1,2,...,\eta$. According to [26], this matrix inequality is equivalent to:

$$\begin{bmatrix} -K_i+N_i(\alpha) & \tilde{M}_i^T(h,\alpha)K_i \\ K_i\tilde{M}_i(h,\alpha) & -K_i \end{bmatrix} < 0 \tag{5.16}$$

Lemma 3 *The matrix P_2^i is positive definite if and only if there does not exist a constant nonzero vector x such that $g(t)x=0$ for all $t\in[0,h]$.*

Proof Consider an arbitrary nonzero vector $x\in\Re^{l_1+l_2+\ldots+l_k}$. Since R_i is positive definite, the term $x^Tg(t)^TR_ig(t)x$ is always nonnegative. Hence, its integral over the interval $[0,h]$ is zero if and only if $g(t)x=0$ for all $t\in[0,h]$. In addition, the term $\bar{M}_i(t)^TQ_i\bar{M}_i(t)$ is always nonnegative due to the positive semi-definiteness of the matrix Q_i. If there exists a nonzero vector x such that $g(t)x=0$ for all $t\in[0,h]$, then it is straightforward to show that the matrix P_2^i is not positive definite, otherwise the term $x^Tg(t)^TR_ig(t)x$ is always positive, which implies the positive definiteness of the matrix P_2^i. ■

Lemma 3 presents a necessary and sufficient condition for the positive definiteness of the matrix P_2^i, which "almost always" holds in practice. It is assumed in the remainder of the chapter that the matrix P_2^i is positive definite, as this assumption is required for the development of the main result.

Theorem 2 *The matrix inequality (5.16) is equivalent to the following matrix inequality:*

$$\begin{bmatrix} \Phi_1^i & (\Phi_2^i)^T & (\Phi_4^i)^T \\ \Phi_2^i & \Phi_3^i & (\Phi_5^i)^T \\ \Phi_4^i & \Phi_5^i & -I \end{bmatrix} < 0 \tag{5.17}$$

where

$$\Phi_1^i := -K_i + \begin{bmatrix} P_0^i + P_1^i\alpha HC_i + (P_1^i\alpha HC_i)^T & P_1^i\alpha G \\ (P_1^i\alpha G)^T & 0 \end{bmatrix},$$

$$\Phi_2^i := K_i\begin{bmatrix} M_i(h) & 0 \\ FC_i & E \end{bmatrix},$$

$$\Phi_3^i := -K_i - K_i\begin{bmatrix} \bar{M}_i(h)(P_2^i)^{-1}\bar{M}_i(h)^T & 0 \\ 0 & 0 \end{bmatrix}K_i,$$

$$\Phi_4^i := (P_2^i)^{\frac{1}{2}}\alpha\begin{bmatrix} HC_i & G \end{bmatrix},$$

$$\Phi_5^i := \begin{bmatrix} (P_2^i)^{-\frac{1}{2}}\bar{M}_i(h)^T & 0 \end{bmatrix}K_i$$

Proof One can write the inequality (5.16) as follows:

$$\begin{bmatrix} \Phi_1^i & (\Phi_2^i)^T \\ \Phi_2^i & \Phi_3^i \end{bmatrix} - \begin{bmatrix} (\Phi_4^i)^T \\ (\Phi_5^i)^T \end{bmatrix} (-I) \begin{bmatrix} \Phi_4^i & \Phi_5^i \end{bmatrix} < 0$$

The matrix inequality (5.17) yields by applying the Schur complement formula to the above inequality. ■

It can be easily verified that in the absence of the block entry Φ_3^i, the matrix given in the left side of (5.17) is in the form of LMI. Moreover, this block entry cannot be converted to the LMI form due to the negative quadratic term inside it [27]. Thus, the technique introduced in [28] will now be used to remedy this drawback. Consider the arbitrary positive definite matrices $\Gamma_1, \Gamma_2, ..., \Gamma_\eta$ with the same dimensions as $K_1, K_2, ..., K_\eta$, respectively. Since P_2^i is assumed to be positive definite, one can write $(K_i - \Gamma_i)\Omega_i(K_i - \Gamma_i) \geq 0,\ i \in \bar{\eta}$, where:

$$\Omega_i := \begin{bmatrix} \bar{M}_i(h)(P_2^i)^{-1}\bar{M}_i(h)^T & 0 \\ 0 & 0 \end{bmatrix}$$

Therefore:

$$-K_i\Omega_iK_i \leq \Pi_i, \quad i \in \bar{\eta} \tag{5.18}$$

where $\Pi_i := \Gamma_i\Omega_i\Gamma_i - K_i\Omega_i\Gamma_i - \Gamma_i\Omega_iK_i,\ i \in \bar{\eta}$.

Theorem 3 *There exist positive definite matrices $K_1, K_2, ..., K_\eta$ satisfying the matrix inequality (5.17) if and only if there exist positive definite matrices $K_1, K_2, ..., K_\eta$ and $\Gamma_1, \Gamma_2, ..., \Gamma_\eta$ satisfying the following matrix inequalities:*

$$\begin{bmatrix} \Phi_1^i & (\Phi_2^i)^T & (\Phi_4^i)^T \\ \Phi_2^i & -K_i + \Pi_i & (\Phi_5^i)^T \\ \Phi_4^i & \Phi_5^i & -I \end{bmatrix} < 0, \quad i \in \bar{\eta} \tag{5.19}$$

Proof If there exist positive definite matrices $K_1, K_2, ..., K_\eta$ and $\Gamma_1, \Gamma_2, ..., \Gamma_\eta$ satisfying (5.19), then according to (5.18), the matrices $K_1, K_2, ..., K_\eta$ satisfy (5.17) as well. On the other hand, suppose that there exist positive definite matrices $K_1, K_2, ..., K_\eta$ satisfying the matrix inequality (5.17). Choosing $\Gamma_i = K_i$ for $i = 1, 2, ..., \eta$, one can easily verify that $\Phi_3^i = -K_i + \Pi_i$. Hence, the inequality (5.17) is equivalent to the inequality (5.19) in this case. ■

It is to be noted that the matrix inequality (5.19) is LMI for the variables K_i, $i \in \bar{\eta}$, and α, if the matrices Γ_i, $i \in \bar{\eta}$, are set to be fixed. The following algorithm is proposed to compute the coefficients $\alpha_1, \alpha_2, ..., \alpha_\eta$ in order to obtain the desired GSHF $f(t)$.

Algorithm 1

Step 1) *Set $\alpha = \breve{\alpha}$ (where $\breve{\alpha}$ is defined in Remark 1) and solve the discrete Lyapunov equation (5.9) in order to obtain $K_1, K_2, ..., K_\eta$.*

Step 2) *Set $\Gamma_i = K_i$ for all $i \in \bar{\eta}$, where the matrices K_i, $i \in \bar{\eta}$, are obtained in Step 1.*

Step 3) *Minimize J given by (5.8) for $K_1, K_2, ..., K_\eta$ and α subject to*

- *The LMI constraint (5.19)*
- *$K_i > 0$ for all $i \in \bar{\eta}$*
- *The constraint that each entry of α whose index belongs to the set* **E** *must be zero.*

Step 4) *If $\sum_{i=1}^{\eta} \|K_i - \Gamma_i\| < \delta$, where δ is a predetermined error margin, go to Step 6.*

Step 5) *Set $\Gamma_i = K_i$ for $i = 1, 2, ..., \eta$, where the matrices K_i, $i \in \bar{\eta}$, are obtained by solving the optimization problem in Step 3. Go to Step 3.*

Step 6) *The value obtained for α is sufficiently close to the optimal value, and substituting the resultant matrices K_i, $i \in \bar{\eta}$, into (5.8) gives the minimum value of J. Note that the coefficients $\alpha_1, \alpha_2, ..., \alpha_k$ can be obtained from (5.5).*

Remark 2 *The matrix $\breve{\alpha}$ defined in Remark 1 has the property that all of the eigenvalues of the matrices $\tilde{M}_i(h, \breve{\alpha})$, $i \in \bar{\eta}$, are located inside the unit circle. As a result, the discrete Lyapunov equation (5.9) is solvable for $\alpha = \breve{\alpha}$, as it is required in Step 1 of the above algorithm.*

Remark 3 *It can be easily verified that the value of J decreases each time that the optimization problem of Step 3 is solved, which indicates that the algorithm is monotone decreasing. On the other hand, since the inequality (5.18) will be converted to the equality if $\Gamma_i = K_i$, Algorithm 1 should ideally stop when $\sum_{i=1}^{\eta} \|K_i - \Gamma_i\| = 0$ in order to obtain the exact result. However, since it is desirable that the algorithm be halted in a finite time, Step 4 is added. It is to be noted that δ determines (indirectly) the closeness of the performance index obtained to its minimum value.*

5.5 Numerical examples

Example 1 This example can be found in [28; 29], and represents the ship-steering system with two distinct modes. Consider two systems with the following parameters:

$$A_1 = \begin{bmatrix} -0.298 & -0.279 & 0 \\ -4.370 & -0.773 & 0 \\ 0 & 1 & 0 \end{bmatrix}, \quad B_1 = \begin{bmatrix} 0.116 \\ -0.773 \\ 0 \end{bmatrix}$$

$$A_2 = \begin{bmatrix} -0.428 & -0.339 & 0 \\ -2.939 & -1.011 & 0 \\ 0 & 1 & 0 \end{bmatrix}, \quad B_2 = \begin{bmatrix} 0.150 \\ -1.011 \\ 0 \end{bmatrix}$$

and $C_1 = C_2 = I$. Assume that the initial state of each of these systems is a random variable whose covariance matrix is equal to the identity matrix, and that $h = 0.1sec$. Assume also that it is desired to find a GSHF which minimizes the performance index J given by (5.4) with $R_i = Q_i = I,\ i = 1,2,$ while it has the following structure:

$$f(t) = \begin{bmatrix} *+*\,sin(t) & * & *+*\,e^{-t} \end{bmatrix} \tag{5.20}$$

where the symbol "*" represents constant values which are to be found. Note that no compensator is considered in this example. The following basis functions and coefficient matrices can therefore be defined for (5.20):

$$f_1(t) = sin(t), \quad f_2(t) = 1, \quad f_3(t) = e^{-t}$$

$$\alpha_1 = \begin{bmatrix} * & 0 & 0 \end{bmatrix}, \quad \alpha_2 = \begin{bmatrix} * & * & * \end{bmatrix}, \quad \alpha_3 = \begin{bmatrix} 0 & 0 & * \end{bmatrix}$$

It is to be noted that the "*" elements used above imply that these entries of the vectors α_1, α_2 and α_3 are the ones that are not set equal to zero. Let Algorithm 1 start with the following initial stabilizing point:

$$\breve{\alpha} = \begin{bmatrix} 0 & 0 & 0 \\ -7 & 5 & 5 \\ 0 & 0 & 0 \end{bmatrix}$$

Accordingly, the optimal value for α will be:

$$\alpha = \begin{bmatrix} 2.805 & 0 & 0 \\ -3.925 & 2.117 & -0.188 \\ 0 & 0 & 1.480 \end{bmatrix}$$

which results in a GSHF equal to the following:

$$\begin{bmatrix} -3.925+2.805\,sin(t) & 2.117 & -0.188+1.480\,e^{-t} \end{bmatrix}$$

The corresponding performance index is 31.581.

Example 2 Consider a two-input two-output system $\mathscr{S}$ consisting of two single-input single-output (SISO) agents with the following state-space matrices:

$$A = \begin{bmatrix} 0.5 & 0 & 0 \\ 0 & -2.5 & 0 \\ 0 & 0 & -5.5 \end{bmatrix}, \quad B = \begin{bmatrix} 0 & -2 \\ -2 & 2 \\ -4 & 0 \end{bmatrix}, \quad C = \begin{bmatrix} -2 & 0 & 2 \\ 0 & -2 & 2 \end{bmatrix}$$

It is desired to design a high-performance decentralized controller with the diagonal information flow structure for this system. It can be easily concluded from [21] that $\lambda = 0.5$ is a decentralized fixed mode (DFM) of the system. Thus, there is no LTI controller to stabilize the system. As a result, the available methods to design a continuous-time LTI controller (e.g., see [27]) are incapable of handling this problem. On the other hand, it results from [30] that $\lambda = 0.5$ is an unstructured DFM (as opposed to a structured one), which implies that this DFM can be eliminated by means of sampling. Choose $h = 1sec$, and denote the discrete-time equivalent model of the system $\mathscr{S}$ with $\mathscr{S}_d$. As expected from the result given in [30], the system $\mathscr{S}_d$ does not have any DFM, and consequently, is decentrally stabilizable. If the algorithm presented in [19] is exploited to design a discrete-time *static* stabilizing controller for the system $\mathscr{S}_d$, it will fail. This signifies that in order to design a decentralized controller for the system $\mathscr{S}$, two different types of controllers can be used: a dynamic discrete-time controller or a periodic controller. These two possibilities are explained in the following:

i) Let a deadbeat dynamic stabilizing controller K_c for the system $\mathscr{S}_d$ be designed by using the method given in [21]. To evaluate the performance of the system $\mathscr{S}$ under the discrete-time controller K_c, consider the performance index (5.4) and assume that $Q = R = I$, and that the initial state of the system $\mathscr{S}$ is a random variable with the identity covariance matrix. In this case, the corresponding performance index will be equal to 83439.49, which is inadmissibly large. The resultant output of the second agent, for instance, is depicted in Figure 5.1 for $x(0) = [0.5\ \ 0.5\ \ 0.5]^T$. This illustrates the inferior intersample ripple, while the magnitude of the output is approximately zero at the sampling points. To improve this ripple effect, a hold controller K_h is desired to be added to the control system. Assume that the hold function $f(t)$ is desirable to have the following form:

$$\mathrm{diag}([* + * \; cos(850t) \quad * + * \; cos(850t)])$$

Using Algorithm 1 with several iterations results in the hold function $f(t) = \mathrm{diag}([1.003 - 0.071cos(850t) \;,\; 0.958 + 0.754cos(850t)])$, and the corresponding performance index turns out to be 81517.97. This indicates an improvement of about 2.36% by using the hold controller K_h. However, this enhancement is not noticeable.

ii) It is desired to find out whether there exists a hold controller K_h to stabilize the system $\mathscr{S}$ by itself (i.e., without any compensator K_c). Consider the following basis functions for the hold function $f(t)$:

$$f_i(t) = u_e\left(t - \frac{i-1}{2}\right) - u_e\left(t - \frac{i}{2}\right), \quad i = 1, 2$$

where $u_e(\cdot)$ denotes the unit-step function. It is to be noted that this GSHF is equivalent to a piecewise constant function with two different levels. Applying the result of [19] to Lemma 1 leads to the controller K_h with the hold function:

$$\mathrm{diag}([-1.4f_1(t) - 0.185f_2(t) \quad 0.5f_1(t) + f_2(t)]) \tag{5.21}$$

The resulting performance index is equal to 2121.18. Hence, Algorithm 1 can now be utilized to adjust the coefficients of this hold function properly. The optimal $f(t)$ obtained will be equal to:

$$\mathrm{diag}([-2.71f_1(t) + 1.08f_2(t) \quad 0.97f_1(t) - 0.30f_2(t)]) \tag{5.22}$$

and the performance index of the closed-loop system will be 301.73. This implies that a high-performance stabilizing controller is designed for the ill-controllable system $\mathscr{S}$. The outputs of the first and the second agents of $\mathscr{S}$ under the hold functions (5.21) and (5.22) are illustrated in Figures 5.2(a) and 5.2(b) for $x(0) = [0.5 \;\; 0.5 \;\; 0.5]^T$. In addition, the inputs of the first and the second agents of $\mathscr{S}$ are depicted in Figures 5.3(a) and 5.3(b). Note that the solid curves and the dotted curves correspond to the GSHFs given in (5.22) and (5.21), respectively. The value of the cost function J is plotted for the first 150 iterations in Figure 5.4, which demonstrates the fast convergence of the algorithm, specially in the beginning. It is to be noted that although the initial point in the optimization algorithm is chosen far from the optimal point, the convergence speed is very good.

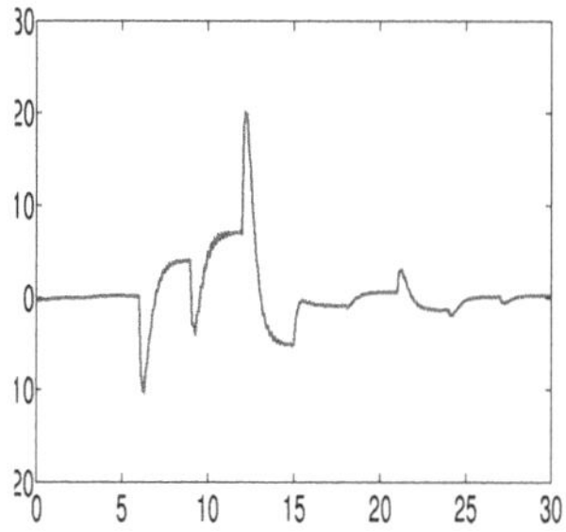

Figure 5.1: The output of the second agent of the system in Example 2 under the dead-beat controller.

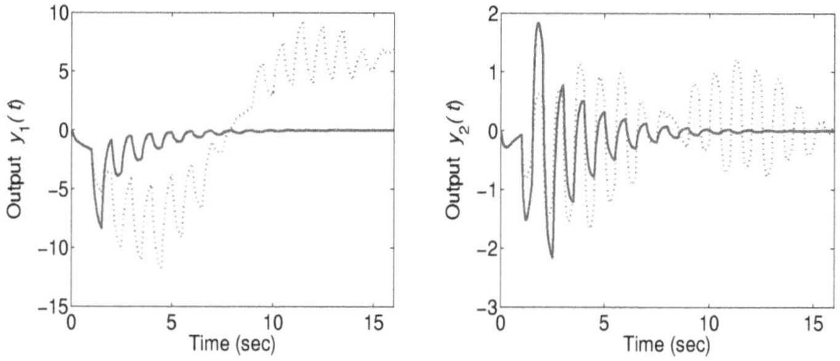

Figure 5.2: The outputs of the first agent and the second agent of the system in Example 2 are depicted in (a) and (b), respectively, under the GSHFs (5.21) (dotted curves), and (5.22) (solid curves).

5.6 Conclusions

In this chapter, a method is proposed to design a decentralized periodic output feedback with a prescribed form, e.g. polynomial, piecewise constant, sinusoidal, etc., to simultaneously stabilize a set of continuous-time LTI systems and minimize a predefined guaranteed continuous-time quadratic performance index, which is, in fact, the summation of the performance indices of all of the systems. The

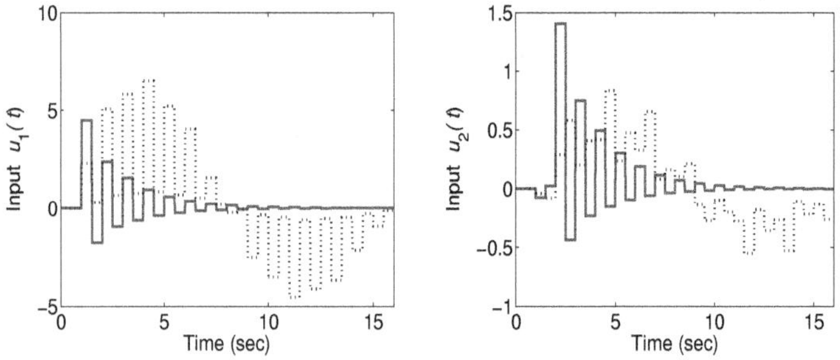

Figure 5.3: The inputs of the first agent and the second agent of the system in Example 2 are depicted in (a) and (b), respectively, under the GSHFs (5.21) (dotted curves), and (5.22) (solid curves).

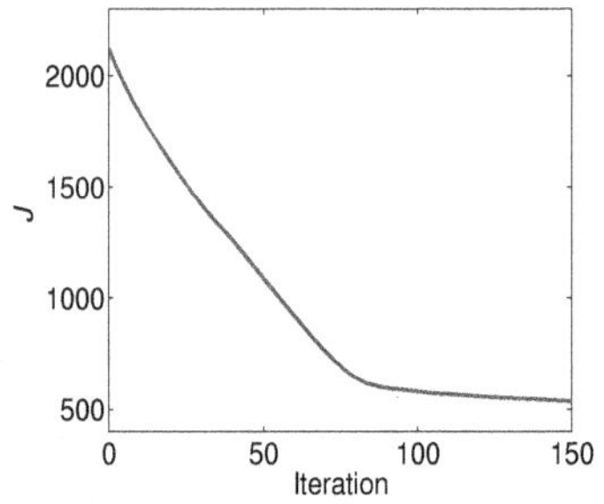

Figure 5.4: The value of the cost function J for the first 150 iterations.

design procedure is accomplished in three phases: First, the problem is formulated as a set of matrix inequalities. Next, it is converted to a set of linear matrix inequalities, which represent necessary and sufficient conditions for the existence of such a structurally constrained controller. An algorithm is then presented to solve the resultant LMI problem. Simulation results demonstrate the effectiveness of the proposed method.

Bibliography

[1] A. B. Chammas and C. T. Leondes, "On the design of linear time-invariant systems by periodic output feedback: Part I, discrete-time pole placement," *International Journal of Control*, vol. 27, pp. 885-894, 1978.

[2] P. T. Kabamba, "Control of linear systems using generalized sampled-data hold functions," *IEEE Transactions on Automatic Control*, vol. 32, no. 9, pp. 772-783, 1987.

[3] Y. C. Juan and P. T. Kabamba, "Optimal hold functions for sampled data regulation," *Automatica*, vol. 27, no. 1, pp. 177-181, 1991.

[4] P. T. Kabamba and C. Yang, "Simultaneous controller design for linear time-invariant systems," *IEEE Transactions on Automatic Control*, vol. 36, no. 1, pp. 106-111, 1991.

[5] A. Feuer and G. C. Goodwin, "Generalized sample hold functions-frequency domain analysis of robustness, sensitivity, and intersample difficulties," *IEEE Transactions on Automatic Control*, vol. 39, no. 5, pp. 1042-1047, 1994.

[6] J. Zhang and C. Zhang, "Robustness analysis of control systems using generalized sample hold functions," *Proceedings of the 33rd IEEE Conference on Decision and Control*, Lake Buena Vista, FL, pp. 237-242, 1994.

[7] M. Rossi and D. E. Miller, "Gain/phase margin improvement using static generalized sampled-data hold functions," *Systems & Control Letters*, vol. 37, no. 3, pp. 163-172, 1999.

[8] C. Fonte, M. Zasadzinski, C. Bernier-Kazantsev, and M. Darouach, "On the simultaneous stabilization of three or more plants," *IEEE Transactions on Automatic Control*, vol. 46, no. 7, pp. 1101-1107, 2001.

[9] G. Fernandez-Anaya, S. Munoz-GutiErrez, R. A. Sanchez-Guzman, and W. Mayol-Cuevas, "Simultaneous stabilization using evolutionary strategies," *International Journal of Control*, vol. 68, no. 6, pp. 1417-1435, 1997.

[10] D. Youla, J. Bongiorno, and C. Lu, "Single-loop feedback stabilization of linear multivariable dynamical plants," *Automatica*, vol. 10, no. 2, pp. 159-173, 1974.

[11] M. Vidyasagar and N. Viswanadham, "Algebraic design techniques for reliable stabilization," *IEEE Transactions on Automatic Control*, vol. 27, no. 5, pp. 1085-1095, 1982.

[12] Y. Jia and J. Ackermann, "Condition and algorithm for simultaneous stabilization of linear plant," *Automatica*, vol. 37, no. 9, pp. 1425-1434, 2001.

[13] V. Blondel and M. Gevers, "Simultaneous stabilizability of three linear systems is rationally undecidable," *Mathematics of Control, Signals, and Systems*, vol. 6, no. 2, pp. 135-145, 1993.

[14] O. Toker and H. Özbay, "On the NP-hardness of solving bilinear matrix inequalities and simultaneous stabilization with static output feedback," *Proceedings of the 1995 American Control Conference*, Seattle, Washington, pp. 2525-2526, 1995.

[15] D. E. Miller and M. Rossi, "Simultaneous stabilization with near optimal LQR performance," *IEEE Transactions on Automatic Control*, vol. 46, no. 10, pp. 1543-1555, 2001.

[16] T. J. Tarn and T. Yang, "Simultaneous stabilization of infinite-dimensional systems with periodic output feedback," *Linear Circuit Systems and Signals Processing: Theory and Application*, pp. 409-424, 1988.

[17] G. L. Hyslop, H. Schattler, and T. J. Tarn, "Descent algorithms for optimal periodic output feedback control," *IEEE Transactions on Automatic Control*, vol. 37, no. 12, pp. 1893-1904, 1992.

[18] H. Werner, "An iterative algorithm for suboptimal periodic output feedback control," *UKACC International Conference on Control*, pp. 814-818, 1996.

[19] Y. Y. Cao and J. Lam, "A computational method for simultaneous LQ optimal control design via piecewise constant output feedbac," *IEEE Transactions on Systems, Man, and Cybernetics*, vol. 31, no. 5, pp. 836-842, 2001.

[20] Y. C. Juan and P. T. Kabamba, "Simultaneous pole assignment in linear periodic systems by constrained structure feedback," *IEEE Transactions on Automatic Control*, vol. 34, no. 2, pp. 168-173, 1989.

[21] E. J. Davison and T. N. Chang, "Decentralized stabilization and pole assignment for general proper systems," *IEEE Transactions on Automatic Control*, vol. 35, no. 6, pp. 652-664, 1990.

[22] A. G. Aghdam, "Decentralized control design using piecewise constant hold functions," *Proceedings of the 2006 American Control Conference*, Minneapolis, Minnesota, 2006.

[23] S. H. Wang, "Stabilization of decentralized control systems via time-varying controllers," *IEEE Transactions on Automatic Control*, vol. 27, no. 3, pp. 741-744, 1982.

[24] J. L. Yanesi and A. G. Aghdam, "Optimal generalized sampled-data hold functions with a constrained structure," *Proceedings of the 2006 American Control Conference*, Minneapolis, Minnesota, 2006.

[25] M. K. Tippett and D. Marchesin, "Upper bounds for the solution of the discrete algebraic Lyapunov equatio," *Automatica*, vol. 35, no. 8, pp. 1485-1489, 1999.

[26] S. Boyd, L. E Ghaoui, E. Feron, and V. Balakrishnan, Linear matrix inequalities in system and control theory, Philadelphia, PA: SIAM, 1994.

[27] Y. Y. Cao, Y. X. Sun,and J. Lam, "Simultaneous stabilization via static output feedback and state feedback," *IEEE Transactions on Automatic Control*, vol. 44, no. 6, pp. 1277-1282, 1999.

[28] Y. Y. Cao and Y. X. Sun, "Static output feedback simultaneous stabilization: ILMI approach," *International Journal of Control*, vol. 70, no. 5, pp. 803-814, 1998.

[29] G. D. Howitt and R. Luus, "Control of a collection of linear systems by linear state feedback control," *International Journal of Control*, vol. 58, pp. 79-96, 1993.

[30] Ü. Özgüner and E. J. Davison, "Sampling and decentralized fixed modes," *Proceedings of the 1985 American Control Conference*, pp. 257-262, 1985.

Chapter 6

Elimination of Fixed Modes by Means of Sampling

6.1 Abstract

This chapter deals with structurally constrained periodic control design for interconnected systems. It is assumed that the system is linear time-invariant (LTI), observable and controllable and that its modes are distinct and nonzero. It is shown that the notions of a quotient fixed mode and a structured decentralized fixed mode are equivalent for this class of systems. If the system is stabilizable under a general decentralized controller (e.g. nonlinear, time-varying), then it is proved that a decentralized LTI discrete-time compensator followed by a zero-order hold can stabilize the system. Moreover, the problem of designing a structurally constrained controller for an interconnected system is converted to the design of a decentralized compensator and a decentralized hold function for an expanded system. In addition, the problem when the structurally constrained hold function is desired to have a special form, e.g. piecewise constant, polynomial, etc., is formulated. A procedure is given in this case to design the optimal hold function with respect to a quadratic performance index.

6.2 Introduction

The notion of a decentralized fixed mode (DFM) is introduced in [1] to identify those modes of an interconnected system, which are fixed with respect to any linear time-invariant (LTI) decentralized control law. In addition, the notion of a structurally fixed mode is introduced in [2], and it is shown that a mode is fixed due to either the structure of the system, or the perfect matching of the parameters. This idea is used in [3] to classify the modes of a decoupled decentralized system as being either structured or unstructured. It is also shown in [3] that the distinct and nonzero unstructured DFMs of a system can be eliminated by means of sampling. Furthermore, the notion of a quotient fixed mode (QFM) is introduced in [4] to determine when an interconnected system is decentrally stabilizable under a general nonlinear and time-varying control law.

On the other hand, there has been a considerable amount of interest in the past several years towards control of continuous-time systems by means of periodic feedback, or so-called generalized sampled-data hold functions (GSHF) [5-11]. Periodic feedback control signal is constructed by sampling the output of the system at equidistant time instants, and multiplying the samples by a continuous-time hold function, which is defined over one sampling interval. Several advantages and disadvantages of GSHF and its application in practical problems have been thoroughly investigated and different design techniques are proposed in the literature [5; 6; 8; 9].

This chapter investigates the decentralized periodic control design problem for the observable and controllable finite dimensional LTI systems with distinct and nonzero modes. It is shown that for this broad class of systems, the notions of QFMs and structured DFMs are identical. Using this result, it is proved that if a system is decentrally stabilizable by means of a general control law, then there exists a decentralized LTI discrete-time controller with a simple zero-order hold (ZOH) to stabilize it. This is due to the elimination of the non-quotient DFMs, in the sampled system. Then, the problem of controlling the aforementioned class of LTI systems by means of a structurally constrained controller is studied. The design of a structurally constrained controller consisting of a discrete-time LTI compensator and a GSHF is converted to the design of a decentralized LTI discrete-time compensator and a decentralized GSHF for an expanded system.

In addition, a method is proposed to design an optimal structurally constrained controller with a certain configuration. This is accomplished by using the expanded system corresponding to the given structural constraint. A decentralized discrete-time compensator with a simple ZOH is designed to stabilize the expanded system. Then, the ZOH in the control system is replaced by a GSHF with a prespecified form, e.g. piecewise constant, polynomial, etc., to improve the performance of the overall system by minimizing a continuous-time quadratic cost function, which accounts for the intersample ripple effect. The significance of the results obtained in this chapter is demonstrated in two numerical examples.

6.3 Preliminaries

Consider a linear time-invariant (LTI) interconnected system $\mathscr{S}$ consisting of ν subsystems $S_1, S_2, ..., S$ represented by:

$$\begin{aligned} \dot{x}(t) &= Ax(t) + \sum_{i=1}^{\nu} B_i u_i(t) \\ y_i(t) &= C_i x(t), \qquad i \in \bar{\nu} := \{1, 2, ..., \nu\} \end{aligned} \tag{6.1}$$

where $x(t) \in \Re^n$ is the state, and $u_i(t) \in \Re^{m_i}$ and $y_i(t) \in \Re^{r_i}$, $i \in \bar{\nu}$, are the input and the output of the ith subsystem, respectively. Assume that the modes of the system $\mathscr{S}$ are all distinct and nonzero, and that the state-space model is in the decoupled form, i.e.,

$$A = \text{diag}([\sigma_1 \ , \ \sigma_2 \ , \ ... \ , \ \sigma_n]) \tag{6.2}$$

where $\sigma_i \neq 0$, $\sigma_i \neq \sigma_j$, $\forall i, j \in \{1, 2, ..., n\}$, $i \neq j$. Assume that the system $\mathscr{S}$ is controllable and observable. Define the following matrices:

$$B := \begin{bmatrix} B_1 & B_2 & \cdots & B_\nu \end{bmatrix}, \quad C := \begin{bmatrix} C_1^T & C_2^T & \cdots & C_\nu^T \end{bmatrix}^T \tag{6.3}$$

Define also,

$$\begin{aligned} &m := \sum_{i=1}^{\nu} m_i, \quad r := \sum_{i=1}^{\nu} r_i, \\ &u(t) := \begin{bmatrix} u_1(t)^T & u_2(t)^T & \cdots & u_\nu(t)^T \end{bmatrix}^T, \quad y(t) := \begin{bmatrix} y_1(t)^T & y_2(t)^T & \cdots & y_\nu(t)^T \end{bmatrix}^T \end{aligned} \tag{6.4}$$

The notion of a decentralized fixed mode (DFM) is defined in [1] and [12] to identify those modes of an interconnected system which remain fixed with respect to any LTI controller with a block diagonal information flow structure. Throughout this chapter, the term "decentralized controller" is referred to the union of local controllers. In order to specify the local subsystems corresponding to the local controllers, the subsystems are enclosed within parentheses throughout the chapter, if necessary. For instance, a decentralized controller for the system $\mathscr{S}(S_1, S_2, S_3)$ is the union of the local controllers $u_i(t) = g_i(y_i(t), t)$, $i \in \{1, 2, 3\}$, corresponding to the subsystems S_1, S_2, S_3, while a decentralized controller for $\mathscr{S}(S_1 \cup S_2, S_3)$ is composed of two local controllers: one for the new subsystem consisting of S_1 and S_2, and the other one for the subsystem S_3.

The following definitions are extracted from [3].

Definition 1 *Assume that $\lambda \in sp(A)$ is a DFM of the system $\mathscr{S}(S_1, S_2, ..., S_\nu)$. λ is defined to be a structured decentralized fixed mode (SDFM) of the system, if it remains a DFM after arbitrary perturbing the nonzero entries of the matrices B and C.*

It is to be noted that in the definition of SDFM given in [3], the nonzero elements of A (the elements on the main diagonal) are also assumed to be perturbed. However, it can be easily verified that such assumption is not necessary, in general.

Definition 2 *Assume that λ is a DFM of the system $\mathscr{S}(S_1, S_2, ..., S_\nu)$. Then, λ is called an unstructured decentralized fixed mode (UDFM) if it is not a SDFM of $\mathscr{S}(S_1, S_2, ..., S_\nu)$, or equivalently, if it is resulted from the exact matching of the nonzero elements of the matrices A, B and C in the state-space model.*

Define $\mathscr{S}_d$ as the discrete-time equivalent model of $\mathscr{S}$ with a constant sampling period $h > 0$ and a ZOH. Hence, the state-space representation of $\mathscr{S}_d$ is as follows:

$$\begin{aligned} x[\kappa+1] &= \bar{A}x[\kappa] + \sum_{i=1}^{\nu} \bar{B}_i u_i[\kappa] \\ y_i[\kappa] &= C_i x[\kappa], \quad i \in \bar{\nu} \end{aligned} \tag{6.5}$$

where the discrete argument corresponding to the samples of any signal is enclosed in brackets (e.g., $x[\kappa] = x(\kappa h)$). It can be easily shown that and $\bar{A} = e^{AT}$, $\bar{B}_i = A^{-1}(\bar{A} - I_n)B_i$, $i \in \bar{\nu}$, where I_n represents the $n \times n$ identity matrix (note that the matrix A is assumed to have no eigenvalues in the origin, and thus, it is nonsingular). Denote the subsystems of $\mathscr{S}_d$ corresponding to $S_1, S_2, ..., S_\nu$ with $\mathscr{S}_{d_1}, \mathscr{S}_{d_2}, ..., \mathscr{S}_{d_\nu}$. The following Lemma is extracted from [3].

Lemma 1 *Assume that the system $\mathscr{S}(S_1, S_2, ..., S_\nu)$ contains $P_u \geq 0$ UDFMs and $P_s \geq 0$ SDFMs λ_i, $i = 1, 2, ..., P_s$. The discrete-time system $\mathscr{S}_d(\mathscr{S}_{d_1}, \mathscr{S}_{d_2}, ..., \mathscr{S}_{d_\nu})$ comprises P_s SDFMs $e^{\lambda_i h}$, $i = 1, 2, ..., P_s$, and no UDFMs for almost all values of h.*

Remark 1 *The term "for almost all values of h" in Lemma 1 means that the values of h for which the discrete-time system $\mathscr{S}_d(\mathscr{S}_{d_1}, \mathscr{S}_{d_2}, ..., \mathscr{S}_{d_\nu})$ has UDFMs, lie on a hypersurface in the one dimensional space [13] (i.e. among infinite possible values for h, only a finite number of them violate Lemma 1).*

Lemma 1 states that the UDFMs of the system $\mathscr{S}(S_1, S_2, ..., S_\nu)$ are eliminated by sampling, while the structured ones cannot be removed. The questions arise as how to identify the type of DFMs, and how to design stabilizing controllers for the systems with SDFMs (if possible). A procedure is proposed in [3] to determine the type of the DFMs of the systems consisting of only two single-input single-output (SISO) subsystems (i.e. $\nu = 2$, $m_1 = m_2 = r_1 = r_2 = 1$).

The notion of a quotient fixed mode (QFM) is introduced in [4] to investigate the stabilizability of interconnected systems under a general decentralized control law (i.e. nonlinear and time-varying). Since the definition of QFM is essential in the development of the main result of this chapter, it is explained in the next two definitions.

Definition 3 *Define the structural graph of the system $\mathscr{S}$ as a digraph with ν vertices which has a directed edge from the ith vertex to the jth vertex if and only if $C_j(sI - A)^{-1}B_i \neq 0$, for any $i, j \in \bar{\nu}$. The structural graph of the system $\mathscr{S}$ is denoted by $\mathscr{G}$.*

Partition $\mathscr{G}$ into the minimum number of strongly connected subgraphs denoted by $G_1, G_2, ..., G_l$ (note that a digraph is called strongly connected iff there exists a directed path from any vertex to any other vertices of the graph [4; 14]). Define the subsystem $\tilde{S}_i$, $i = 1, 2, ..., l$, as the union of all subsystems of $\mathscr{S}$ corresponding to the vertices in the subgraph G_i (note that vertex i in the graph $\mathscr{G}$ represents the subsystem S_i, for any $i \in \bar{\nu}$).

Definition 4 *λ is said to be a QFM of the system $\mathscr{S}(S_1, S_2, ..., S_\nu)$, if it is a DFM of the system $\mathscr{S}(\tilde{S}_1, \tilde{S}_2, ..., \tilde{S}_l)$.*

Lemma 2 *The system $\mathscr{S}(S_1, S_2, ..., S_\nu)$ is stabilizable under a general decentralized controller, if and only if it does not have any QFM in the closed right-half complex plane.*

The above Lemma is given in [4], where it is also stated that a candidate stabilizing decentralized controller can be a time-varying control law, or a vibrational one. In the next section, it will be shown that the notions of SDFMs and QFMs are identical.

6.4 Effect of sampling on DFM

Notation 1 *For any $i \in \bar{\nu}$:*

- *Denote the* (j_1, j_2) *entry of* B_i *with* $b_i^{j_1,j_2}$*, for any* $1 \leq j_1 \leq n,\ 1 \leq j_2 \leq m_i$.

- *Denote the* (j_1, j_2) *entry of* C_i *with* $c_i^{j_1,j_2}$*, for any* $1 \leq j_1 \leq r_i,\ 1 \leq j_2 \leq n$.

Theorem 1 σ_i*,* $i \in \{1,2,...,n\}$*, is a SDFM of the system* $\mathscr{S}(S_1, S_2, ..., S_\nu)$*,* $\nu \geq 2$*, if and only if there exist a permutation of* $\{1,2,...,\nu\}$ *denoted by distinct integers* $i_1, i_2, ..., i_\nu$ *and an integer* p *between 1 and* $\nu - 1$ *such that* $b_{j_1}^{i,\alpha} = c_{j_2}^{\beta,i} = 0$ *and* $b_{j_1}^{\mu,\alpha} c_{j_2}^{\beta,\mu} = 0$*, for all* j_1, j_2, α, β *and* μ *given by*

$$\begin{aligned} & j_1 \in \{i_1, i_2, ..., i_p\},\ \ j_2 \in \{i_{p+1}, i_{p+2}, ..., i_\nu\},\ \ 1 \leq \alpha \leq m_{j_1}, \\ & 1 \leq \beta \leq r_{j_2},\ \ 1 \leq \mu \leq n,\ \ \mu \neq i \end{aligned} \tag{6.6}$$

Proof It is known that σ_i is a DFM of the system $\mathscr{S}(S_1, S_2, ..., S_\nu)$ if and only if there exist a permutation of $\{1,2,...,\nu\}$ denoted by distinct integers $i_1, i_2, ..., i_\nu$ and an integer p between 0 and ν such that the rank of the following matrix is less than n [12]:

$$\begin{bmatrix} A - \sigma_i I_n & B_{i_1} & B_{i_2} & \dots & B_{i_p} \\ C_{i_{p+1}} & 0 & 0 & \dots & 0 \\ C_{i_{p+2}} & 0 & 0 & \dots & 0 \\ \vdots & \vdots & \vdots & \ddots & \vdots \\ C_{i_\nu} & 0 & 0 & \dots & 0 \end{bmatrix} \tag{6.7}$$

(note that 0 in the above matrix represents a zero block matrix with proper dimension). In addition, since it is assumed that the system $\mathscr{S}$ is controllable as well as observable, the rank of the matrix (6.7) is equal to n for $p = 0$ and $p = \nu$. Therefore, the condition $0 \leq p \leq \nu$ given above can be replaced by $1 \leq p \leq \nu - 1$. It is clear that the rank of the matrix $A - \sigma_i I_n$ is $n - 1$, and also, the ith column and the ith row of this matrix are both zeros. Hence, if there exists a nonzero entry either in the ith column or in the ith row of the matrix given in (6.7), its rank will be at least n. As a result, the rank of the matrix in (6.7) is less than n, if and only if both of the following conditions hold:

i) All of the entries of the ith column and the ith row of the matrix given in (6.7) are zero, i.e. $b_{j_1}^{i,\alpha} = c_{j_2}^{\beta,i} = 0$ for any α, β, j_1, and j_2 satisfying (6.6).

ii) The rank of the following matrix (which is a sub-matrix of the one given by (6.7)) is less than n:

$$\begin{bmatrix} \sigma_1^i & \dots & 0 & 0 & \dots & 0 & b_{j_1}^{1,\alpha} \\ \vdots & \ddots & \vdots & \vdots & \ddots & \vdots & \vdots \\ 0 & \dots & \sigma_{i-1}^i & 0 & \dots & 0 & b_{j_1}^{i-1,\alpha} \\ 0 & \dots & 0 & \sigma_{i+1}^i & \dots & 0 & b_{j_1}^{i+1,\alpha} \\ \vdots & \ddots & \vdots & \vdots & \ddots & \vdots & \vdots \\ 0 & \dots & 0 & 0 & \dots & \sigma_n^i & b_{j_1}^{n,\alpha} \\ c_{j_2}^{\beta,1} & \dots & c_{j_2}^{\beta,i-1} & c_{j_2}^{\beta,i+1} & \dots & c_{j_2}^{\beta,n} & 0 \end{bmatrix} \tag{6.8}$$

for any α, β, j_1, and j_2 satisfying (6.6), where $\sigma_j^i := \sigma_j - \sigma_i$, $i, j \in \{1, 2, ..., n\}$. Partition the matrix given by (6.8) into four sub-matrices, and denote it with:

$$\begin{bmatrix} A_i & \Phi_1 \\ \Phi_2 & 0 \end{bmatrix} \tag{6.9}$$

where $A_i \in \Re^{(n-1)\times(n-1)}$, $\Phi_1 \in \Re^{(n-1)\times 1}$, and $\Phi_2 \in \Re^{1\times(n-1)}$. Since the matrix A_i is nonsingular (because it is assumed that $\sigma_1, ..., \sigma_n$ are distinct), one can write:

$$\det \begin{bmatrix} A_i & \Phi_1 \\ \Phi_2 & 0 \end{bmatrix} = -\det(A_i) \times \det(\Phi_2 A_i^{-1} \Phi_1) \tag{6.10}$$

Thus, the rank of the matrix given in (6.8) is less than n, if and only if the scalar $\Phi_2 A_i^{-1} \Phi_1$ is equal to 0.

It can be concluded from the above discussion, that σ_i is a SDFM, if and only if the condition (i) and the equality

$$\sum_{\mu=1,\ \mu\neq i}^{n} \frac{\bar{b}_{j_1}^{\mu,\alpha} \bar{c}_{j_2}^{\beta,\mu}}{\sigma_\mu^i} = 0 \tag{6.11}$$

both hold, where $\bar{b}_{j_1}^{\mu,\alpha}$ and $\bar{c}_{j_2}^{\beta,\mu}$ represent any arbitrary nonzero multiples of $b_{j_1}^{\mu,\alpha}$ and $c_{j_2}^{\beta,\mu}$, respectively, for $\mu = 1, 2, ..., n$. This condition is equivalent to the equality $b_{j_1}^{\mu,\alpha} c_{j_2}^{\beta,\mu} = 0$ for $\mu = 1, 2, ..., n,\ \mu \neq i$. ■

Corollary 1 *Assume that σ_i, $i \in \{1, 2, ..., n\}$, is a SDFM of the system $\mathscr{S}(S_1, S_2, ..., S_\nu)$. There exist a permutation of $\{1, 2, ..., \nu\}$ denoted by distinct integers $i_1, i_2, ..., i_\nu$ and an integer p between 1 and $\nu - 1$, such that*

i) The following two matrices are not full-rank:

$$\begin{bmatrix} A-\sigma_i I_n & B_{i_1} & B_{i_2} & \dots & B_{i_p} \end{bmatrix}, \quad \begin{bmatrix} A-\sigma_i I_n & C_{i_{p+1}}^T & C_{i_{p+2}}^T & \cdots & C_{i_v}^T \end{bmatrix}^T \tag{6.12}$$

ii) The following equality holds for any complex number $s \neq \sigma_j$, $j=1,2,...,n$:

$$\begin{bmatrix} C_{i_{p+1}}^T & C_{i_{p+2}}^T & \cdots & C_{i_v}^T \end{bmatrix}^T (A-sI_n)^{-1} \begin{bmatrix} B_{i_1} & B_{i_2} & \dots & B_{i_p} \end{bmatrix} = 0 \tag{6.13}$$

Proof The proof is straightforward and follows directly from Theorem 1. ■

Theorem 2 *The SDFMs of the system $\mathscr{S}(S_1,S_2,...,S_v)$ are identical to its QFMs.*

Proof Assume that σ_i is a SDFM of the system $\mathscr{S}(S_1,S_2,...,S_v)$, and consider the integers $i_1,i_2,...,i_v$ in Corollary 1. Define now two new composite subsystems $\mathbf{S}_1$ and $\mathbf{S}_2$, where $\mathbf{S}_1$ is composed of p subsystems $S_1,S_2,...,S_p$, and $\mathbf{S}_2$ is composed of $v-p$ subsystems $S_{p+1},S_{p+2},...,S_v$. One can easily conclude from (6.7) and the characteristics of DFM given in [12], that σ_i is a DFM of the system $\mathscr{S}(\mathbf{S}_1,\mathbf{S}_2)$.

On the other hand, condition (ii) of Corollary 1 implies that the transfer function matrix form $\mathbf{S}_1$ to $\mathbf{S}_2$ is zero, i.e., the system $\mathscr{S}$ consisting of the two subsystems $\mathbf{S}_1$ and $\mathbf{S}_2$ is not strongly connected (note that a system consisting of two subsystems is strongly connected if and only if the transfer function from each of its subsystems to the other one is nonzero). Furthermore, since the system $\mathscr{S}$ has already been broken down into the subsystems $\tilde{S}_1,\tilde{S}_2,...,\tilde{S}_l$ (where l denotes the minimum number of strongly connected subgraphs of $\mathscr{G}$, as discussed earlier), which are not strongly connected to each other, it can be easily verified that there exist a permutation of $\{1,2,...,l\}$ denoted by distinct integers $j_1,j_2,...,j_l$, and a number ζ such that $\mathbf{S}_1=\tilde{S}_{j_1}\cup\tilde{S}_{j_2}\cup\cdots\cup\tilde{S}_{j_\zeta}$ and $\mathbf{S}_2=\tilde{S}_{j_{\zeta+1}}\cup\tilde{S}_{j_{\zeta+2}}\cup\cdots\cup\tilde{S}_{j_l}$. This implies that any DFM of the system $\mathscr{S}(\mathbf{S}_1,\mathbf{S}_2)$ is also a DFM of the system $\mathscr{S}(\tilde{S}_1,\tilde{S}_2,...,\tilde{S}_l)$ (because the decentralized control structure for $S(\mathbf{S}_1,\mathbf{S}_2)$ includes the decentralized control structure for $\mathscr{S}(\tilde{S}_1,\tilde{S}_2,...,\tilde{S}_l)$). Thus, since it is proved that σ_i is a DFM of the system $\mathscr{S}(\mathbf{S}_1,\mathbf{S}_2)$, it is a DFM of the system $\mathscr{S}(\tilde{S}_1,\tilde{S}_2,...,\tilde{S}_l)$ as well. On the other hand, it is known from Definition 4, that the DFMs of the system $\mathscr{S}(\tilde{S}_1,\tilde{S}_2,...,\tilde{S}_l)$ are equivalent to the QFMs of the system $\mathscr{S}(S_1,S_2,...,S_v)$. Therefore, σ_i is a QFM of the system $\mathscr{S}(S_1,S_2,...,S_v)$.

Assume now that λ is a QFM of the system $\mathscr{S}(S_1,S_2,...,S_v)$. Hence, λ is either a SDFM or an UDFM. If it is an UDFM, then it follows from Lemma 1 that λ is not fixed with respect to a discrete-time controller and a ZOH. A well-known property of QFM [3], however, is that λ is fixed with respect to any type of control law [4], which contradicts the original assumption. This implies that λ is a SDFM of the system $\mathscr{S}(S_1, S_2, ..., S_\nu)$, and this completes the proof. ∎

Remark 2 *It can be concluded from conditions (i) and (ii) in Corollary 1 and the discussion in the proof of Theorem 2, that if σ_i is a SDFM of the system $\mathscr{S}(S_1, S_2, ..., S_\nu)$ (or equivalently a QFM), then the system can be partitioned into two subsystems $\mathbf{S}_1$ and $\mathbf{S}_2$, such that σ_i is an uncontrollable mode of the system $\mathscr{S}$ from the input of the subsystem $\mathbf{S}_1$, and an unobservable mode of $\mathscr{S}$ from the output of the subsystem $\mathbf{S}_2$. Furthermore, the transfer function matrix from the input of $\mathbf{S}_1$ to the output of $\mathbf{S}_2$ is zero.*

Remark 3 *Assume that the system $\mathscr{S}(S_1, S_2, ..., S_\nu)$ contains the DFMs λ_i, $i = 1, 2, ..., P_s$, which are also QFMs, and the DFMs $\bar{\lambda}_i$, $i = 1, 2, ..., P_u$, which are not QFMs. It follows from Theorem 2 and Lemma 6.1, that the discrete-time equivalent model $\mathscr{S}_d(\mathscr{S}_{d_1}, \mathscr{S}_{d_2}, ..., \mathscr{S}_{d_\nu})$ has only the DFMs $e^{\lambda_i h}$, $i = 1, 2, ..., P_s$, which corresponds to the QFMs of $\mathscr{S}(S_1, S_2, ..., S_\nu)$, for almost all values of h. In other words, the DFMs $\bar{\lambda}_i$, $i = 1, 2, ..., P_u$, of $\mathscr{S}(S_1, S_2, ..., S_\nu)$ will be eliminated by sampling.*

This means that if the system $\mathscr{S}(S_1, S_2, ..., S_\nu)$ is decentrally stabilizable, then there exists a sampled-data decentralized controller to stabilize it. It is shown in [4] and [14], that a system with no unstable QFMs can be stabilized by an appropriate time-varying control law. However, the implementation of a sampled-data controller is simpler in general, and has its unique advantages. Structurally constrained control of systems with stable QFMs using sampled-data hold functions will be spelled out in the next section.

6.5 Constrained generalized sampled-data hold controller

In this section, a new compelling reason for the effectiveness of GSHF will be presented and some of its properties will be studied.

Assume that the structure of the overall controller for the system $\mathscr{S}$ has some prespecified constraints [7]. These constraints determine which outputs y_i $(i \in \bar{\nu})$ are available to construct any specific input u_j $(j \in \bar{\nu})$ of the system. In order to simplify the problem formulation for the control constraint, a $\nu \times \nu$ block matrix $\mathscr{K}$ with binary entries is defined, where its (i, j) block entry, $i, j \in \bar{\nu}$, is a $m_i \times r_j$

matrix with all entries equal to 1 if the output of the jth subsystem can contribute to the construction of the input of the ith subsystem, and is a $m_i \times r_j$ zero matrix otherwise. The matrix $\mathscr{K}$ represents the control constraint, and will be referred to as the information flow matrix. In the special case, when the entries of the matrix $\mathscr{K}$ are all equal to 1, the corresponding controller is centralized, and when $\mathscr{K}$ is block diagonal, the corresponding controller is decentralized.

Consider the following discrete-time compensator K_c for the system $\mathscr{S}$:

$$\begin{aligned} z[\kappa+1] &= Ez[\kappa] + Fy[\kappa] \\ \phi[\kappa] &= Gz[\kappa] + Hy[\kappa] \end{aligned} \tag{6.14}$$

and also, the hold controller K_h:

$$u(t) = F(t)\phi[\kappa], \quad \kappa h \leq t < (\kappa+1)h, \quad \kappa = 0, 1, 2, \ldots \tag{6.15}$$

where $F(t) = F(t+h)$, $\forall t \geq 0$. Note that the matrices E, F, G, H, and the function $F(t)$ are desired to be designed such that the overall controller consisting of K_c and K_h meet the design specifications. Assume now that an information flow matrix $\mathscr{K}$ is given for the system $\mathscr{S}$. In order to design the compensator K_c and the hold controller K_h such that the overall control structure complies with the information flow matrix $\mathscr{K}$, a block-diagonal (decentralized) structure is assumed for the compensator K_c in [7], while the structure of the hold controller K_h is assumed to meet the control constraint inferred from $\mathscr{K}$. It is evident that this assumption is ill-posed, because the number of the parameters of K_h to be designed is often much less than that of K_c, and also K_c has a significant role in stabilizing the system $\mathscr{S}$ (see the proof of Theorem 4). Hence, it is hereafter assumed that K_h is desired to be decentralized, while the structure of K_c complies with the constraint given by the information flow matrix $\mathscr{K}$. The following procedure is used to form the transfer function matrix $K_c(z) := G(zI_n - E)^{-1}F + H$ of the compensator K_c in order to comply with the information flow matrix $\mathscr{K}$.

Procedure 1 *Replace the (i, j) block entry of $\mathscr{K}$, $i, j \in \bar{\nu}$, with $K_{ij}(z) \in \Re^{m_i \times r_j}$ (whose parameters are yet to be designed) if it is not a zero matrix, i.e., if the output of the jth subsystem can contribute to the construction of the input of the ith subsystem. Denote the resultant matrix with $K_c(z)$. Note that the nonzero block entries of $K_c(z)$ are unknown so far, and are desired to be found so that the closed-loop system satisfies the design specifications.*

The following procedure is used to construct a decentralized LTI compensator $\bar{K}_c$ with the block-diagonal transfer function matrix $\bar{K}_c(z)$. It will be shown how the entries of the desired constrained compensator K_c can be mapped into $\bar{K}_c$.

Procedure 2 *Form $\bar{K}_c(z) \in \Re^{m\times\bar{r}}$ as a block diagonal matrix, whose (i,i) block entry, $i \in \bar{\nu}$, is a $m_i \times \bar{r}_i$ matrix which is obtained from the ith block row of the matrix $K_c(z)$ by contracting it, i.e. by eliminating all of its zero block entries and placing the remaining block entries next to each other, and in the same order that they appear in the corresponding block row of $K_c(z)$.*

As an example, assume that the matrix $K_c(z)$ is as follows:

$$K_c(z) = \begin{bmatrix} K_{11}(z) & 0 & K_{13}(z) & 0 \\ 0 & K_{22}(z) & 0 & 0 \\ 0 & 0 & K_{33}(z) & 0 \end{bmatrix} \tag{6.16}$$

The corresponding matrix $\bar{K}_c(z)$ is obtained to be:

$$\bar{K}_c(z) = \begin{bmatrix} K_{11}(z) & K_{13}(z) & 0 & 0 \\ 0 & 0 & K_{22}(z) & 0 \\ 0 & 0 & 0 & K_{33}(z) \end{bmatrix} \tag{6.17}$$

Let the state-space representation of the decentralized compensator $\bar{K}_c$, whose transfer function matrix is formed in Procedure 2, be denoted by:

$$\begin{aligned} \bar{z}[\kappa+1] &= \bar{E}\bar{z}[\kappa] + \bar{F}\bar{y}[\kappa] \\ \bar{\phi}[\kappa] &= \bar{G}\bar{z}[\kappa] + \bar{H}\bar{y}[\kappa] \end{aligned} \tag{6.18}$$

Assume throughout the chapter, zero initial states for the compensators, i.e., $z[0] = \bar{z}[0] = 0$. Note that since the parameters of the compensator K_c are still unknown at this point, and since $\bar{K}_c$ is formed based on K_c, the parameters $\bar{E}, \bar{F}, \bar{G}$ and $\bar{H}$ are unknown too.

Remark 4 *It can be easily verified that there exists an onto mapping between the nonzero block entries of the matrix $\bar{K}_c(z)$ obtained in Procedure 2, and those of the matrix $K_c(z)$.*

Lemma 3 *For any given $K_c(z)$, there exists a constant matrix T such that $K_c(z) = \bar{K}_c(z)T$, where T is obtained from the matrix $\mathscr{K}$.*

Proof The proof is omitted and may be found in [15]. A procedure is also given in [15] to obtain the transformation matrix T. ■

Define $\bar{\mathscr{S}}$ as an interconnected system consisting of the subsystems $\bar{\mathscr{S}}_1, \bar{\mathscr{S}}_2, \ldots, \bar{\mathscr{S}}_\nu$ with the following state-space representation:

$$\begin{aligned} \dot{\bar{x}}(t) &= A\bar{x}(t) + \sum_{i=1}^{\nu} B_i \bar{u}_i(t) \\ \bar{y}_i(t) &= \bar{C}_i \bar{x}(t), \quad i \in \bar{\nu} \end{aligned} \tag{6.19}$$

where $\bar{C}_i \in \Re^{\bar{r}_i \times n}$, $i \in \bar{\nu}$, and:

$$\left[\begin{array}{cccc} \bar{C}_1^T & \bar{C}_2^T & \cdots & \bar{C}_\nu^T \end{array} \right]^T = TC \tag{6.20}$$

and $\bar{u}_i(t)$ and $\bar{y}_i(t)$ are the input and the output of the subsystem $\bar{\mathscr{S}}_i$. Define now,

$$\begin{aligned} \bar{u}(t) &:= \left[\begin{array}{cccc} \bar{u}_1(t)^T & \bar{u}_2(t)^T & \cdots & \bar{u}_\nu(t)^T \end{array} \right]^T \\ \bar{y}(t) &:= \left[\begin{array}{cccc} \bar{y}_1(t)^T & \bar{y}_2(t)^T & \cdots & \bar{y}_\nu(t)^T \end{array} \right]^T \end{aligned} \tag{6.21}$$

Define also the hold controller $\bar{K}_h$ as:

$$\bar{u}(t) = F(t)\bar{\phi}[\kappa], \quad \kappa h \leq t < (\kappa+1)h, \quad \kappa = 0, 1, 2, \ldots \tag{6.22}$$

Theorem 3 *For any given compensator $K_c(z)$ (or equivalently, any given matrices E, F, G, and H) corresponding to the information flow matrix $\mathscr{K}$, and the hold function $F(t)$, construct the matrix $\bar{K}_c(z)$ by using Procedure 2. The state and the input of the system $\mathscr{S}$ under the compensator $K_c(z)$ and the hold controller K_h are equivalent to those of the system $\bar{\mathscr{S}}$ under the compensator $\bar{K}_c(z)$ and the hold controller $\bar{K}_h$, provided $x(0) = \bar{x}(0)$.*

Proof It is desired first to show that $x(t) = \bar{x}(t)$ and $u(t) = \bar{u}(t)$ for any $0 \leq t < h$. Since $x(0) = \bar{x}(0)$, one can easily conclude that $\bar{y}[0] = Ty[0]$. On the other hand, it follows from Lemma 3, that the transfer functions of the compensators K_c and $\bar{K}_c$ satisfy the equation $K(z) = \bar{K}(z)T$. Note that the inputs of these compensators are $y[k]$ and $\bar{y}[k]$, respectively. Hence, it can be easily concluded that the outputs of these compensators at time $k = 0$ are equal, i.e. $\phi[0] = \bar{\phi}[0]$ (note that $z[0] = \bar{z}[0] = 0$). Thus, the equations (6.15) and (6.22) result in the equality $u(t) = \bar{u}(t)$ for all $t \in [0, h)$. Consequently, one can conclude from the state-space equations of the systems $\mathscr{S}$ and $\bar{\mathscr{S}}$, and the equality $x(0) = \bar{x}(0)$, that $x(t) = \bar{x}(t)$ for all $t \in [0,h)$. Since the states $x(t)$ and $\bar{x}(t)$ are continuous functions of time, $x(h) = \bar{x}(h)$ or equivalently $x[1] = \bar{x}[1]$. Now, one can start from the equality $x[1] = \bar{x}[1]$, and use a similar argument to conclude that $x(t) = \bar{x}(t)$ and $u(t) = \bar{u}(t)$ for all $t \in [h, 2h)$. Continuing this argument will lead to the equalities $x(t) = \bar{x}(t)$ and $u(t) = \bar{u}(t)$, for all $t \in [ih, (i+1)h)$, $\forall i \geq 0$. ■

Remark 5 *Theorem 3 states that instead of designing a structurally constrained compensator K_c and a decentralized hold controller K_h for the system $\mathscr{S}$ to achieve any desired objective (stability, pole placement, etc.), one can equivalently design a decentralized compensator $\bar{K}_c$ and a decentralized hold controller $\bar{K}_h$ for the system $\bar{\mathscr{S}}$. Then, the original compensator K_c can be obtained by using the equation $K_c(z) = \bar{K}_c(z)T$. In addition, the hold function $F(t)$ designed for $\bar{\mathscr{S}}$ can be equivalently considered for $\mathscr{S}$ (because of the relations (6.15) and (6.22)). However, the advantage of this indirect design procedure is that the compensator $\bar{K}_c$ is decentralized (i.e. it has a block diagonal information flow structure). It is to be noted that the decentralized control design problem has been investigated in the literature intensively, and a number of methods are available [7; 12; 10].*

Theorem 4 *If there exist no decentralized compensator $\bar{K}_c$ and decentralized hold controller $\bar{K}_h$ to stabilize the system $\bar{\mathscr{S}}(\bar{\mathscr{S}}_1, \bar{\mathscr{S}}_2, ..., \bar{\mathscr{S}}_\nu)$, then $\bar{\mathscr{S}}(\bar{\mathscr{S}}_1, \bar{\mathscr{S}}_2, ..., \bar{\mathscr{S}}_\nu)$ is not stabilizable under any type of decentralized control law (i.e. nonlinear, time-varying, etc.).*

Proof If the system $\bar{\mathscr{S}}(\bar{\mathscr{S}}_1, \bar{\mathscr{S}}_2, ..., \bar{\mathscr{S}}_\nu)$ has an unstable QFM, it is not decentrally stabilizable according to Lemma 2. If, however, it has no unstable QFM, one can conclude from Remark 3 that there exists a discrete-time decentralized controller to stabilize the discrete-time equivalent model of the system $\bar{\mathscr{S}}(\bar{\mathscr{S}}_1, \bar{\mathscr{S}}_2, ..., \bar{\mathscr{S}}_\nu)$ for almost all sampling periods $h > 0$. Let this discrete-time controller be denoted by $\bar{K}_c$ and the hold function be equal to $\bar{I}$, where $\bar{I} \in \Re^{m \times m}$ is a block diagonal matrix whose (i,i) block entry is a $m_i \times m_i$ matrix with the entries all equal to 1 for any $i \in \bar{\nu}$. In this case, the compensator $\bar{K}_c$ is, in fact, the stabilizing discrete-time controller and the hold function $F(t)$ is a simple ZOH. ■

It can be concluded from Theorem 4 that in the special case, when $\mathscr{K}$ is block diagonal, or equivalently $\mathscr{S} = \bar{\mathscr{S}}$, then a stabilizing compensator K_c is guaranteed to exist for the system $\mathscr{S}$, if and only if the system is decentrally stabilizable. This discussion and the result of Theorem 4 demonstrate the significance of using a compensator in the system to be controlled. The question may arise as why a hold controller is added to the system when the compensator by itself can stabilize it. To answer this question, assume that the non-quotient DFMs of the system $\bar{\mathscr{S}}$ are aimed to be placed at some arbitrary locations. It is pointed out in [7] that there exist infinite candidates for $\bar{K}_h$ to achieve this (because $\bar{K}_c$ by itself can carry out the pole placement). This implies that $\bar{K}_h$ can be designed in such a way that it not only results in the desired pole placement (along with $\bar{K}_c$), but also minimizes a continuous-time performance index to reduce the intersample effect, or even simultaneously stabilize a set of systems. In other words, $\bar{K}_h$ introduces a new set of parameters to the design problem, which can be significantly beneficial to solve a multi-faceted problem.

Remark 6 *Assume that it is desired now to design a decentralized hold controller K_h and a structurally constrained compensator K_c for the system $\mathscr{S}$, such that the following LQR performance index is minimized:*

$$J = \int_0^\infty \left(x(t)^T Q x(t) + u(t)^T R u(t)\right) dt \tag{6.23}$$

where $R \in \Re^{m\times m}$ and $Q \in \Re^{n\times n}$ are positive definite and positive semi-definite matrices, respectively. For simplicity and without loss of generality, assume that Q and R are symmetric. It can be concluded from Theorem 3 and Remark 5 that this problem is equivalent to the problem of designing a decentralized compensator $\bar{K}_c$ and a decentralized hold controller $\bar{K}_h$ for the system $\bar{\mathscr{S}}$, such that the following LQR performance index is minimized:

$$J = \int_0^\infty \left(\bar{x}(t)^T Q \bar{x}(t) + \bar{u}(t)^T R \bar{u}(t)\right) dt \tag{6.24}$$

6.5.1 High-performance structurally constrained controller

According to Theorem 4, if the system $\bar{\mathscr{S}}(\bar{\mathscr{S}}_1, \bar{\mathscr{S}}_2, ..., \bar{\mathscr{S}}_\nu)$ is decentrally stabilizable, then there exists a decentralized discrete-time compensator $\bar{K}_c$ to stabilize the system with the hold function $F(t) = \bar{I}$. Note that the stabilizing compensator $\bar{K}_c$ can be obtained by using any existing method such as pole placement, to achieve any given design specifications. It is desired now to replace the ZOH (i.e., $F(t) = \bar{I}$) with a more advanced hold function, such that the performance of the composite system consisting of $\bar{\mathscr{S}}$ and the stabilizing compensator $\bar{K}_c$ is improved.

On the other hand, it is often advantageous to design a hold function which has a prespecified form, such as piecewise constant, polynomial, etc. [10; 11]. Therefore, assume that the following set of basis functions is considered for the hold function $F(t)$:

$$\mathbf{f} := \{F_1(t), F_2(t), ... F_k(t)\} \tag{6.25}$$

where $F_i(t) \in \Re^{m \times v_i}$, $i = 2, ..., k$, are arbitrary matrix functions and $F_1(t) = \bar{I}$. Thus, $F(t)$ can be written as a linear combination of the basis functions in $\mathbf{f}$ in the following form:

$$F(t) = F_1(t)\alpha_1 + F_2(t)\alpha_2 + \cdots + F_k(t)\alpha_k \tag{6.26}$$

where $\alpha_i \in \Re^{v_i \times m}$, $i = 1, 2, ..., k$, are matrices with certain zero elements, which reflect the structural constraint of $F(t)$. The objective here is to obtain the matrices α_i, $i = 1, 2, ..., k$, to minimize the performance index (6.24) (this will be clarified in Example 2). Note that the first basis $F_1(t)$ is assumed to be equal to $\bar{I}$, because in that case there exists at least one hold function of the form given in (6.26), which along with the compensator $\bar{K}_c$ stabilize the system $\bar{\mathscr{S}}$ (i.e., when $\alpha_1 = I$, $\alpha_i = 0$, $i = 2, 3, ..., k$). It is to be noted that since (6.24) is a continuous-time performance index, it takes the intersample ripple effect into account [11]. The equation (6.26) can be written as $F(t) = W(t)\alpha$, where:

$$\alpha := \left[\alpha_1^T, \alpha_2^T, ..., \alpha_k^T\right]^T, \quad W(t) := [F_1(t), F_2(t), ..., F_k(t)] \tag{6.27}$$

Since some of the entries of the unknown matrices $\alpha_1, ..., \alpha_k$ are set to zero, the matrix α has a spacial structure. To formulate the structure of α, define a set $\mathbf{E}$ which contains all of the indices of the zero entries of α_i for any $i \in \{1, 2, ..., k\}$.

It is known that:

$$\bar{x}(t) = e^{(t-\kappa h)A}\bar{x}(\kappa h) + \int_{\kappa h}^{t} e^{(t-\tau)A}B\bar{u}(\tau)d\tau \tag{6.28}$$

for any $\kappa h \leq t \leq (\kappa+1)h$, $\kappa \geq 0$. Now, let the following matrices be defined:

$$M(t) = e^{tA}, \quad \bar{M}(t) = \int_0^t e^{(t-\tau)A}BW(\tau)d\tau \tag{6.29}$$

Therefore,

$$\bar{x}(t) = M(t-\kappa h)\bar{x}[\kappa] + \bar{M}(t-\kappa h)\alpha\bar{\phi}[k] \tag{6.30}$$

for any $\kappa h \leq t \leq (\kappa+1)h$. It can be easily concluded from (6.19), (6.20), (6.18), and (6.30) by substituting $t = (\kappa+1)h$, that

$$\mathbf{x}[\kappa+1] = \tilde{M}(h, \alpha)\mathbf{x}[\kappa] \tag{6.31}$$

for any $\kappa \geq 0$, where $\mathbf{x}[\kappa] = \left[\begin{array}{cc} \bar{x}[\kappa]^T & \bar{z}[\kappa]^T \end{array} \right]^T$, and

$$\tilde{M}(h,\alpha) := \left[\begin{array}{cc} M(h) + \bar{M}(h)\alpha\bar{H}TC & \bar{M}(h)\alpha\bar{G} \\ \bar{F}TC & \bar{E} \end{array} \right] \tag{6.32}$$

It is straightforward to show that (by using the equation (6.31)):

$$\mathbf{x}[\kappa] = \left(\tilde{M}(h,\alpha)\right)^{\kappa}\mathbf{x}[0], \quad \kappa = 0,1,2,... \tag{6.33}$$

Define now the following matrices:

$$P_0 := \int_0^h \left(M(t)^T Q M(t)\right) dt \tag{6.34a}$$

$$P_1 := \int_0^h \left(M(t)^T Q \bar{M}(t)\right) dt \tag{6.34b}$$

$$P_2 := \int_0^h \left(\bar{M}(t)^T Q \bar{M}(t) + W(t)^T R W(t)\right) dt \tag{6.34c}$$

$$q_0(\alpha) := P_0 + P_1\alpha\bar{H}TC + (P_1\alpha\bar{H}TC)^T + (\alpha\bar{H}TC)^T P_2(\alpha\bar{H}TC) \tag{6.34d}$$

$$q_1(\alpha) := P_1\alpha\bar{G} + (\alpha\bar{H}TC)^T P_2\alpha\bar{G} \tag{6.34e}$$

$$P(\alpha) := \left[\begin{array}{cc} q_0(\alpha) & q_1(\alpha) \\ q_1(\alpha)^T & \bar{G}^T\alpha^T P_2\alpha\bar{G} \end{array} \right] \tag{6.34f}$$

Lemma 4 *For a given α, suppose that the system $\bar{\mathscr{S}}$ is stable under the pair $\bar{K}_c$ and $\bar{K}_h$. The performance index J defined in (6.24) can be written as $J = \mathbf{x}^T(0)\mathbf{K}\mathbf{x}(0)$, where $\mathbf{K}$ satisfies the following discrete Lyapunov equation:*

$$\tilde{M}^T(h,\alpha)\mathbf{K}\tilde{M}(h,\alpha) - \mathbf{K} + P(\alpha) = 0 \tag{6.35}$$

Proof Substituting (6.22) and (6.30) into (6.24) and using (6.31), the performance index can be written as follows:

$$\begin{aligned} J &= \sum_{\kappa=0}^{\infty}\left(\int_{\kappa h}^{(\kappa+1)h}\left(\bar{x}^T(t)Q\bar{x}(t) + \bar{u}^T(t)R\bar{u}(t)\right)dt\right) \\ &= \sum_{\kappa=0}^{\infty}\left(\bar{x}[\kappa]^T P_0\bar{x}[\kappa] + \bar{x}[\kappa]^T P_1\alpha\bar{\phi}[\kappa] + \bar{\phi}[\kappa]^T\alpha^T P_1^T\bar{x}[\kappa] + \bar{\phi}[\kappa]^T\alpha^T P_2\alpha\bar{\phi}[\kappa]\right) \\ &= \sum_{\kappa=0}^{\infty}\bar{\mathbf{x}}[\kappa]^T P(\alpha)\bar{\mathbf{x}}[\kappa] \\ &= \bar{\mathbf{x}}(0)^T\sum_{\kappa=0}^{\infty}\left(\tilde{M}^T(h,\alpha)^{\kappa}P(\alpha)\tilde{M}(h,\alpha)^{\kappa}\right)\bar{\mathbf{x}}(0) \end{aligned} \tag{6.36}$$

As pointed out in the proof of Lemma 1 in [10], since the closed-loop system is stable, all of the eigenvalues of the matrix $\tilde{M}(h,\alpha)$ are located inside the unit circle in the complex plane. Thus, the infinite series

$$\sum_{\kappa=0}^{\infty} (\tilde{M}^T(h,\alpha)^{\kappa} P(\alpha) \tilde{M}(h,\alpha)^{\kappa}) \tag{6.37}$$

converges to $\mathbf{K}$, the solution of the discrete Lyapunov equation (6.35). ■

Remark 7 *The optimization problem is now converted to the problem of minimizing* $\mathbf{x}^T(0)\mathbf{K}\mathbf{x}(0)$ *for the variable* α*, subject to the following constraints:*

1. *Any entry of* α *whose index belongs to the set* $\mathbf{E}$*, must be equal to zero,*

2. $\mathbf{K}$ *satisfies the discrete Lyapunov equation (6.35).*

Since the matrix function $\tilde{M}(h,\alpha)$ *is linear and the matrix function* $P(\alpha)$ *is quadratic with respect to the variable* α*, this optimization problem is the same as the ones solved in [10; 11]. Hence, one can exploit a slight variation of the approach given in [10] to reformulate the problem in the linear matrix inequality (LMI) framework. Note that both of the algorithms presented in [10; 11] require an initial point for* α *such that the corresponding closed-loop system is stable. Therefore, as discussed earlier, one can consider the initial point* $\begin{bmatrix} I & 0 & \cdots & 0 \end{bmatrix}$ *for* α*.*

6.6 Illustrative examples

Example 1 consider the system $\mathscr{S}$ consisting of three SISO subsystems with the following state-space matrices:

$$A = \begin{bmatrix} 1 & 0 & 0 \\ 0 & -2 & 0 \\ 0 & 0 & -3 \end{bmatrix}, \quad B_1 = \begin{bmatrix} 0 \\ 0 \\ -1 \end{bmatrix}, \quad B_2 = \begin{bmatrix} 1 \\ 1 \\ 2 \end{bmatrix}, \quad B_3 = \begin{bmatrix} 2 \\ 1 \\ 5 \end{bmatrix},$$
$$C_1 = \begin{bmatrix} 5 & 3 & 2 \end{bmatrix}, \quad C_2 = \begin{bmatrix} 0 & -1 & 0 \end{bmatrix}, \quad C_3 = \begin{bmatrix} 0 & -2 & 0 \end{bmatrix} \tag{6.38}$$

It can be easily concluded from Theorem 1 (by considering $i_j = j,\ j = 1,2,3$), that $\lambda = 1$ is a SDFM of the system $\mathscr{S}(S_1,S_2,S_3)$. In other words, if the nonzero entries of the vectors $B_i, C_i,\ i = 1,2,3$ are replaced by any arbitrary numbers, then $\lambda = 1$ still remains a DFM of the resultant system.

It is desired now to obtain the QFM of $\mathscr{S}(S_1,S_2,S_3)$. One can easily verify that the structural graph of the system $\mathscr{S}$ is composed of two strongly connected subgraphs corresponding to vertex 1 (as the first subgraph), and vertices 2, 3 (as the second subgraph). Hence, the new subsystem $\tilde{S}_1$ is defined to be the subsystem S_1, and $\tilde{S}_2$ is defined to be the union of S_2 and S_3. Moreover, it can be easily verified that the system $\mathscr{S}(\tilde{S}_1,\tilde{S}_2)$ has a DFM at $\lambda = 1$. Thus, Definition 4 yields that $\lambda = 1$ is a QFM of the system $\mathscr{S}(S_1,S_2,S_3)$. In other words, $\lambda = 1$ is a SDFM as well as a QFM of the system $\mathscr{S}(S_1,S_2,S_3)$. This is in accordance with the result of Theorem 2. It is to be noted that -2 and -3 are not DFMs of the system $\mathscr{S}(S_1,S_2,S_3)$.

Now, let the vectors B_1, C_2, and C_3 in (6.38) be replaced by the following:

$$B_1 = \begin{bmatrix} 0 \\ 3 \\ -4 \end{bmatrix}, \quad C_2 = \begin{bmatrix} 0 \\ -1 \\ -1 \end{bmatrix}^T, \quad C_3 = \begin{bmatrix} 0 \\ -2 \\ -2 \end{bmatrix}^T \tag{6.39}$$

One can easily verify that, in this case, $\lambda = 1$ is a DFM of the system $\mathscr{S}(S_1,S_2,S_3)$, but it is not a QFM; hence, it can be eliminated by means of sampling according to Remark 3. For instance, assume that $h = 1\,\text{sec}$. It is straightforward to show that the modes of the open-loop discrete-time equivalent model are $0.0498, 0.1353, 2.7183$, while those of the closed-loop discrete-time model corresponding to a decentralized feedback with unity gains are $2.0685 \pm 0.7942i, -4.9743$. Since these two sets of modes are disjoint, it can be concluded that the discrete-time equivalent model does not have any DFM, as expected from Remark 3.

Example 2 Consider the system $\mathscr{S}$ consisting of two SISO subsystems with the following state-space matrices:

$$A = \begin{bmatrix} 1 & 0 \\ 0 & -3 \end{bmatrix}, \quad B = \begin{bmatrix} 1 & 2 \\ 1 & 1 \end{bmatrix}, \quad C = \begin{bmatrix} 1 & -2 \\ 1 & 1 \end{bmatrix} \tag{6.40}$$

It is desired to design a stabilizing controller with the information flow matrix $\mathscr{K} = \begin{bmatrix} 0 & 1 \\ 1 & 0 \end{bmatrix}$ for the system $\mathscr{S}$. To achieve this, a compensator K_c and a hold controller K_h will be employed with the sampling period equal to $0.1\,\text{sec}$. The transfer functions $K_c(z)$ and $\bar{K}_c(z)$ have the following structures (by using Procedures 1 and 2):

$$K_c(z) = \begin{bmatrix} 0 & K_{12}(z) \\ K_{21}(z) & 0 \end{bmatrix}, \quad \bar{K}_c(z) = \begin{bmatrix} K_{12}(z) & 0 \\ 0 & K_{21}(z) \end{bmatrix} \tag{6.41}$$

In this simple case, the matrix T introduced in Lemma 3 is found to be $T = \begin{bmatrix} 0 & 1 \\ 1 & 0 \end{bmatrix}$. Accordingly, the system $\bar{\mathscr{S}}(\bar{\mathscr{S}}_1, \bar{\mathscr{S}}_2)$ can be obtained from the equation (6.19). It is straightforward to show that the system $\bar{\mathscr{S}}(\bar{\mathscr{S}}_1, \bar{\mathscr{S}}_2)$ does not have any QFM. Thus, it follows from Theorem 4 that there exists a discrete-time decentralized compensator $\bar{K}_c$ to stabilize the system $\bar{\mathscr{S}}(\bar{\mathscr{S}}_1, \bar{\mathscr{S}}_2)$ along with a simple ZOH. It can be easily verified that the static controller $\bar{K}_c(z) = \begin{bmatrix} 1 & 0 \\ 0 & -1 \end{bmatrix}$ will achieve this. Note that this translates to the following static gain for the original system:

$$K_c(z) = \bar{K}_c(z)T = \begin{bmatrix} 0 & 1 \\ -1 & 0 \end{bmatrix} \tag{6.42}$$

Now, a decentralized hold controller $\bar{K}_h$ is to be designed for the composite system consisting of the system $\bar{\mathscr{S}}(\bar{\mathscr{S}}_1, \bar{\mathscr{S}}_2)$ and the decentralized compensator $\bar{K}_c$ given above. Consider the performance index (6.24), and assume that $Q = R = I_2$. Assume also that $x(0) = \begin{bmatrix} 1 & 1 \end{bmatrix}^T$. If the hold controller $\bar{K}_h$ is a simple ZOH, this performance index will be equal to 0.9390. Suppose now that instead of a simple ZOH, the following basis functions for the hold function $F(t)$ are given:

$$F_1(t) = I_2, \quad F_2(t) = \begin{bmatrix} \sin(t) & 0 \\ 0 & 0 \end{bmatrix} \tag{6.43}$$

The decentralized constraint of the hold controller requires that the coefficients α_1 and α_2 in (6.26) have the following structures:

$$\alpha_1 = \begin{bmatrix} * & 0 \\ 0 & * \end{bmatrix}, \quad \alpha_2 = \begin{bmatrix} * & 0 \\ 0 & 0 \end{bmatrix} \tag{6.44}$$

where $*$ represents the nonzero entries to be determined. The problem of finding the constrained coefficients α_1 and α_2 such that the performance index (6.24) is minimized is discussed in Remark 7. As pointed out there, the optimization problem can be solved by using the LMI approach presented in [10] with the starting point $\begin{bmatrix} I_2 & 0_{2\times 2} \end{bmatrix}$ for the variable α (note that $0_{2\times 2}$ represents a 2×2 zero matrix). This approach results in the following optimal hold function $F(t)$:

$$F(t) = \begin{bmatrix} 0.5594 - 1.7508\sin(t) & 0 \\ 0 & 0.8297 \end{bmatrix} \tag{6.45}$$

The corresponding performance index will be equal to 0.8450. This implies that the hold function given above will improve the performance of the control system by about 12%. Note that for this example, the minimum achievable performance index resulted by using a centralized LQR controller (assuming that all state variables are available in the output) is equal to 0.7818.

6.7 Conclusions

This chapter deals with a broad class of interconnected systems with a constrained control structure. It is proved that the two notions of a structured decentralized fixed mode and a quotient fixed mode in the literature are identical for linear time-invariant, controllable and observable systems with distinct and nonzero eigenvalues. Furthermore, it is shown that if there exists a decentralized controller with a general structure (e.g. nonlinear, time-varying) to stabilize a system belonging to the aforementioned class, then there exists a decentralized LTI discrete-time compensator (with a zero-order hold), which stabilizes the system. Moreover, it is shown that the problem of designing a structurally constrained controller for an interconnected system in order to achieve some design objectives, such as desired pole locations, is equivalent to the problem of designing a decentralized compensator and a decentralized hold controller for the expanded system, to attain the same objectives. In addition, the problem of designing a stabilizing high-performance controller consisting of a decentralized compensator and a decentralized hold controller, where the hold controller is desired to have a special form, e.g. piecewise constant, polynomial, etc., is investigated. The numerical results obtained, demonstrate the effectiveness of the proposed work.

Bibliography

[1] S. H. Wang and E. J. Davison, "On the stabilization of decentralized control systems," *IEEE Trans. Automat. Contrl.*, vol. 18, no. 5, pp. 473-478, Oct. 1973.

[2] M. E. Sezer and D. D. Šiljak, "Structurally fixed modes," *Sys. and Contr. Lett.*, vol. 1, no. 1, pp. 60-64, July 1981.

[3] Ü. Özgüner and E. J. Davison, "Sampling and decentralized fixed modes," *in Proc. 1985 American Contrl. Conf.*, pp. 257-262, 1985.

[4] Z. Gong and M. Aldeen, "Stabilization of decentralized control systems," *Journal of Mathematical Systems, Estimation, and Control*, vol. 7, no. 1, pp. 1-16, 1997.

[5] J. Zhang and C. Zhang, "Robustness analysis of control systems using generalized sample hold functions," *in Proc. 33rd IEEE Conf. on Decision and Contr.*, vol. 1, pp. 237-242, Lake Buena Vista, FL, 1994.

[6] M. Rossi, D. E. Miller, "Gain/phase margin improvement using static generalized sampled-data hold functions," *Sys. and Contr. Lett.*, vol. 37, no. 3, pp. 163-172, 1999.

[7] Y. C. Juan and P. T. Kabamba, "Simultaneous pole assignment in linear periodic systems by constrained structure feedback," *IEEE Trans. Automat. Contrl.*, vol. 34, no. 2, pp. 168-173, Feb. 1989.

[8] Y. C. Juan and P. T. Kabamba, "Optimal hold functions for sampled data regulation," *Automatica*, vol. 27. no. 1, pp. 177-181, Jan. 1991.

[9] A. Feuer and G. C. Goodwin, "Generalized sample hold functionsfrequency domain analysis of robustness, sensitivity, and intersample difficulties," *IEEE Trans. Automat. Contrl.*, vol. 39, no. 5, pp. 1042- 1047, May 1994.

[10] J. L. Yanesi and A. G. Aghdam, "High-performance simultaneous stabilizing periodic feedback control with a constrained structure," *in Proc. 2006 American Contrl. Conf.*, Minneapolis, MN, 2006.

[11] J. L. Yanesi and A. G. Aghdam, "Optimal generalized sampled-data hold functions with a constrained structure," *in Proc. 2006 American Contrl. Conf.*, Minneapolis, MN, 2006.

[12] E. J. Davison and T. N. Chang, "Decentralized stabilization and pole assignment for general proper systems," *IEEE Trans. Automat. Contrl.*, vol. 35, no. 6, pp. 652-664, June 1990.

[13] E. J. Davison and S. H. Wang, "Properties of linear time-invariant multivariable systems subject to arbitrary output and state feedback," *IEEE Trans. on Automat. Contrl.*, vol. 18 , no.1, pp. 24-32, Feb. 1973.

[14] B. Anderson and J. Moore, "Time-varying feedback laws for decentralized control," *IEEE Trans. Automat. Contrl.*, vol. 26, no. 5, pp. 1133-1139, Oct. 1981.

[15] J. Lavaei and A. G. Aghdam, "A necessary and sufficient condition for the existence of a LTI stabilizing decentralized overlapping controller," *in Proc. of 45th IEEE Conf. on Decision and Contrl.*, San Diego, CA, 2006.

Chapter 7

Characterization of Decentralized and Quotient Fixed Modes Via Graph Theory

7.1 Abstract

This chapter deals with the decentralized control of systems with distinct modes. A simple graph-theoretic approach is first proposed to identify those modes of the system which cannot be moved by means of a linear time-invariant decentralized controller. To this end, the system is transformed into its Jordan state-space representation. Then, a matrix is computed, which has the same order as the transfer function matrix of the system. A bipartite graph is constructed from the computed matrix. Now, the problem of characterizing the decentralized fixed modes of the system reduces to verifying if this graph has a complete bipartite subgraph with a certain property. Analogously, a graph-theoretic method is presented to compute the modes of the system which are fixed with respect to any general (nonlinear and time-varying) decentralized controller. The proposed approaches are quite simpler than the existing ones.

7.2 Introduction

Many real-world systems can be envisaged as the interconnected systems consisting of a number of subsystems. Normally, the desirable control structure for this class of systems is decentralized, which comprises a set of local controllers for the subsystems [1; 2; 3; 4; 5; 6]. Decentralized control theory has found applications in large space structure, communication networks, power systems, etc. [7; 8; 9; 10]. More recently, simultaneous stabilization of a set of decentralized systems and decentralized periodic control design are investigated in [11; 12].

The notion of a decentralized fixed mode (DFM) was introduced in [1], where it was shown that any mode of a system which is not a DFM can be placed freely in the complex plane by means of an appropriate linear time-invariant (LTI) controller. An algebraic characterization of DFMs was presented in [13]. A method was then proposed in [14] to characterize the DFMs of a system in terms of its transfer function. It was shown in [15] that the DFMs of any system can be attained by computing the transmission zeros of a set of systems derived from the original system. In [2], an algorithm was presented to identify the DFMs of the system by checking the rank of a set of matrices. It is worth noting that the number of the systems whose transmission zeros need to be checked in [15] and the number of matrices whose ranks are to be computed in [2] depend exponentially on the number of the subsystems of the original system. This means that while these methods are theoretically developed for any multi-input multi-output (MIMO) system, they are computationally ill-conditioned. The method introduced in [16] addresses this shortcoming by partitioning the system into a number of modified subsystems, obtained based on the strong connectivity of the system's graph. Then, instead of finding the DFMs of the original system, one can compute the DFMs of the modified subsystems to reduce the corresponding computational complexity. However, the computational burden can still be high when the system consists of several strongly connected subsystems. In general, the method given in [16] is more effective for medium-sized systems, while the one in [2] is only appropriate for small-sized systems. It is to be noted that the method introduced in [2] is widely used in the literature for the characterization of the DFMs.

On the other hand, the notion of a quotient fixed mode (QFM) is introduced in [17] to identify those modes of the system which are fixed with respect to general (nonlinear and time-varying) decentralized controllers. The properties of QFM is further investigated in [3], where it is asserted that the non-quotient DFMs of a broad class of systems can by eliminated by means of sampling.

This chapter aims to present simple approaches to find the DFMs and the QFMs of a system with distinct modes. To this end, a matrix is obtained first, which resembles the transfer function matrix of the system at one point. Then, a bipartite graph is constructed in terms of this matrix. It is shown that having a complete bipartite subgraph with a certain property is equivalent to having a DFM. A similar method is pursued to obtain the QFMs of the system. The combinatorial approaches proposed in the present chapter are substantially simpler than the conventional methods for finding the DFMs and QFMs. The efficacy of the proposed methods is demonstrated in two numerical examples.

7.3 Preliminaries

Consider a LTI interconnected system $\mathscr{S}$ consisting of ν subsystems $S_1, S_2, ..., S_\nu$, represented by:

$$\begin{aligned} \dot{x}(t) &= Ax(t) + \sum_{j=1}^{\nu} B_j u_j(t) \\ y_i(t) &= C_i x(t) + \sum_{j=1}^{\nu} D_{ij} u_j(t), \quad i \in \bar{\nu} := \{1, 2, ..., \nu\} \end{aligned} \tag{7.1}$$

where $x(t) \in \Re^n$ is the state, and $u_i(t) \in \Re^{m_i}$ and $y_i(t) \in \Re^{r_i}$, $i \in \bar{\nu}$, are the input and the output of the ith subsystem, respectively. Suppose the eigenvalues of A are distinct. Write the matrix A as $T\mathbf{A}T^{-1}$, where T is the eigenvector matrix of A. Denote the matrix $\mathbf{A}$ as follows:

$$\mathbf{A} = \begin{bmatrix} \sigma_1 & 0 & \cdots & 0 \\ 0 & \sigma_2 & \cdots & 0 \\ \vdots & \cdots & \ddots & \vdots \\ 0 & 0 & \cdots & \sigma_n \end{bmatrix} \tag{7.2}$$

where σ_i, $i \in \{1, 2, ..., n\}$ denote the modes of the system $\mathscr{S}$. Therefore, the system $\mathscr{S}$ can be represented in the decoupled form as:

$$\begin{aligned} \dot{\mathbf{x}}(t) &= \mathbf{A}\mathbf{x}(t) + \sum_{j=1}^{\nu} \mathbf{B}_j u_j(t) \\ y_i(t) &= \mathbf{C}_i \mathbf{x}(t) + \sum_{j=1}^{\nu} \mathbf{D}_{ij} u_j(t), \quad i \in \bar{\nu} \end{aligned} \tag{7.3}$$

where $\mathbf{D}_{ij} = D_{ij}$, $i, j \in \bar{\nu}$ and

$$\begin{bmatrix} \mathbf{B}_1 & \cdots & \mathbf{B}_\nu \end{bmatrix} = T^{-1} \begin{bmatrix} B_1 & \cdots & B_\nu \end{bmatrix}, \; \begin{bmatrix} \mathbf{C}_1 & \cdots & \mathbf{C}_\nu \end{bmatrix} = \begin{bmatrix} C_1 & \cdots & C_\nu \end{bmatrix} T \tag{7.4}$$

Throughout this chapter, the term "decentralized controller" is referred to the union of all local controllers. In order to specify the local subsystems associated with the local controllers, the subsystems are enclosed within parentheses throughout the chapter, if necessary. For instance, a decentralized controller for the system $\mathscr{S}(S_1, S_2, S_3)$ is the union of the local controllers $u_i(t) = g_i(y_i(t), t)$, $i \in \{1, 2, 3\}$, corresponding to the subsystems S_1, S_2, S_3. Some of the important notions for different types of fixed modes will be given next, which are essential for the main results of the chapter.

Definition 1 [1] *$\lambda \in sp(A)$ is said to be a decentralized fixed mode (DFM) of the system $\mathscr{S}$, if it remains a mode of the closed-loop system under any arbitrary decentralized static feedback. In other words, $\lambda \in sp(A)$ is a DFM of the system $\mathscr{S}$ if:*

$$\lambda \in sp\left(A + \sum_{i=1}^{\nu} B_i K_i C_i\right), \quad \forall\, K_i \in \Re^{m_i \times r_i},\ i \in \bar{\nu} \tag{7.5}$$

It can be shown that a DFM is fixed with respect to any arbitrary dynamic LTI decentralized controller. However, it is interesting to note that a proper non-LTI controller can eliminate certain types of DFMs [3]. In other words, a DFM is not necessarily fixed with respect to a time-varying or nonlinear control structure.

Definition 2 *Define the structural graph of the system $\mathscr{S}$ as a digraph with ν vertices which has a directed edge from the ith vertex to the jth vertex if $C_j(sI - A)^{-1}B_i \neq 0$, for any $i, j \in \bar{\nu}$. The structural graph of the system $\mathscr{S}$ is denoted by $\mathscr{G}$.*

Partition $\mathscr{G}$ into the minimum number of strongly connected subgraphs denoted by $G_1, G_2, ..., G_l$ (note that a digraph is called strongly connected iff there exists a directed path from any vertex to any other vertices of the graph [16; 3]). Define the subsystem $\tilde{S}_i$, $i = 1, 2, ..., l$, as the union of all subsystems of $\mathscr{S}$ corresponding to the vertices in the subgraph G_i (note that vertex j in the graph $\mathscr{G}$ represents the subsystem S_j, for any $j \in \bar{\nu}$).

Definition 3 [16] *Assume that the system $\mathscr{S}$ is strictly proper, i.e. $D = 0$. The mode λ is said to be a QFM of the system $\mathscr{S}(S_1, S_2, ..., S_\nu)$, if it is a DFM of the system $\mathscr{S}(\tilde{S}_1, \tilde{S}_2, ..., \tilde{S}_l)$.*

It can be shown in [16] that a QFM is fixed with respect to any arbitrary (nonlinear or time-varying) decentralized controller.

In order to clarify the definition of a QFM, consider the system $\mathscr{S}$ with the parameters given below:

$$\begin{aligned}
&\mathbf{A} = \ \mathrm{diag}([1\ ,\ -2\ ,\ -3]),\\
&\mathbf{B}_1 = [0\ \ 0\ -1]^T,\ \mathbf{B}_2 = [1\ \ 1\ \ 2]^T,\ \mathbf{B}_3 = [2\ \ 1\ \ 5]^T,\\
&\mathbf{C}_1 = [5\ \ 3\ \ 2],\ \mathbf{C}_2 = [0\ -1\ \ 0],\ \ \mathbf{C}_3 = [0\ -2\ \ 0]
\end{aligned} \tag{7.6}$$

The transfer function matrix of this system will be equal to:

$$\mathbf{C}(sI-\mathbf{A})^{-1}\mathbf{B} = \begin{bmatrix} -2s-6 & 12s+13 & 23s+26 \\ 0 & -s-2 & -s-2 \\ 0 & -2s-4 & -2s-4 \end{bmatrix} \tag{7.7}$$

Hence, the structural graph of the system $\mathscr{S}$ is composed of two strongly connected subgraphs corresponding to vertex 1 (as the first subgraph), and vertices 2, 3 (as the second subgraph). Therefore, the new subsystem $\tilde{S}_1$ is defined to be the subsystem S_1, and $\tilde{S}_2$ is defined to be the union of S_2 and S_3. The union of the subsystems $\tilde{S}_1$ and $\tilde{S}_2$ is sometimes referred to as a quotient system [16]. It is important to note that:

- The DFMs of the system $\mathscr{S}(S_1, S_2, S_3)$ are also the modes of the closed-loop system in Figure 7.1, for any arbitrary dynamic LTI controllers K_1, K_2 and K_3. It can be easily verified that $\lambda = 1$ is the only DFM of the system $\mathscr{S}$ given by (7.6).

- The QFMs of the system $\mathscr{S}(S_1, S_2, S_3)$ are defined to be the DFMs of the system $\mathscr{S}(\tilde{S}_1, \tilde{S}_2)$, i.e., the fixed modes of the closed-loop system shown in Figure 7.2, for any arbitrary LTI controllers $\tilde{K}_1$ and $\tilde{K}_2$. For instance, it is easy to show that $\lambda = 1$ is a QFM of the system $\mathscr{S}$ given by (7.6).

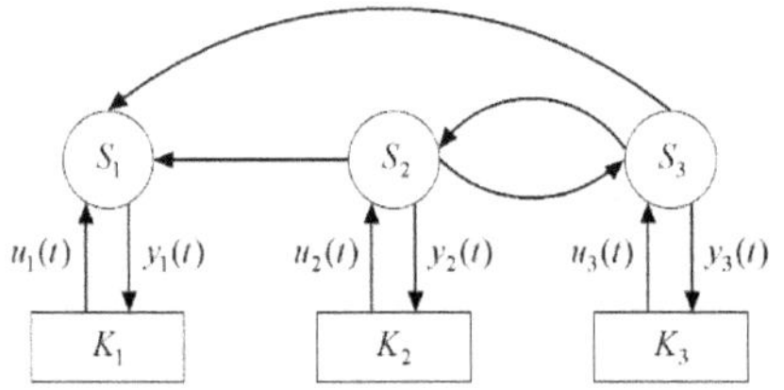

Figure 7.1: The schematic of the decentralized control system $\mathscr{S}$ used for obtaining the DFMs.

7.4 Characterization of decentralized fixed modes

It is desired in this section to present a simple procedure to obtain the DFMs of the system $\mathscr{S}$.

Notation 1 *For any* $i, j \in \bar{\mathcal{V}}$:

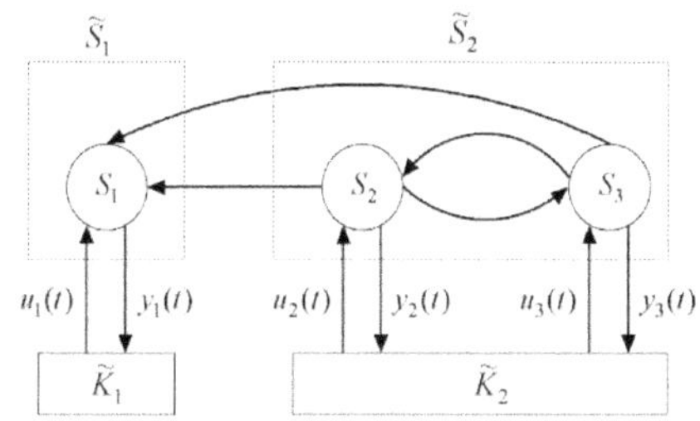

Figure 7.2: The schematic of the decentralized control system $\mathscr{S}$ used for obtaining the QFMs.

- *Denote the* (μ_1, μ_2) *entry of* $\mathbf{B}_i$ *with* $\mathbf{b}_i^{\mu_1,\mu_2}$, *for any* $1 \leq \mu_1 \leq n$, $1 \leq \mu_2 \leq m_i$.
- *Denote the* (μ_1, μ_2) *entry of* $\mathbf{C}_i$ *with* $\mathbf{c}_i^{\mu_1,\mu_2}$, *for any* $1 \leq \mu_1 \leq r_i$, $1 \leq \mu_2 \leq n$.
- *Denote the* (μ_1, μ_2) *entry of* $\mathbf{D}_{ij}$ *with* $\mathbf{d}_{ij}^{\mu_1,\mu_2}$, *for any* $1 \leq \mu_1 \leq r_i$, $1 \leq \mu_2 \leq m_j$.

The following theorem formulates the DFMs of the system $\mathscr{S}$.

Theorem 1 *Assume that the mode* σ_i, $i \in \{1,2,...,n\}$, *is controllable as well as observable.* σ_i *is a DFM of the system* $\mathscr{S}$, $\nu \geq 2$, *if and only if there exist a permutation of* $\{1,2,...,\nu\}$ *denoted by distinct integers* $i_1, i_2, ..., i_\nu$ *and an integer* $p \in [1, \nu-1]$ *such that* $\mathbf{b}_\eta^{i,\alpha} = \mathbf{c}_\gamma^{\beta,i} = 0$, *and:*

$$\sum_{\mu=1,\ \mu\neq i}^{n} \frac{\mathbf{b}_\eta^{\mu,\alpha}\mathbf{c}_\gamma^{\beta,\mu}}{\sigma_\mu - \sigma_i} = \mathbf{d}_{\gamma\eta}^{\beta,\alpha} \tag{7.8}$$

for all η, γ, α *and* β *given by:*

$$\eta \in \{i_1, i_2, ..., i_p\},\ \ \gamma \in \{i_{p+1}, i_{p+2}, ..., i_\nu\},\ \ 1 \leq \alpha \leq m_\eta,\ \ 1 \leq \beta \leq r_\gamma \tag{7.9}$$

Proof It is known that σ_i is a DFM of the system $\mathscr{S}(S_1, S_2, ..., S_\nu)$ if and only if there exist a permutation of $\{1,2,...,\nu\}$ denoted by distinct integers $i_1, i_2, ..., i_\nu$ and an integer $p \in [0, \nu]$ such that the rank of the following matrix is less than n [2]:

$$\begin{bmatrix} \mathbf{A} - \sigma_i I_n & \mathbf{B}_{i_1} & \mathbf{B}_{i_2} & \cdots & \mathbf{B}_{i_p} \\ \mathbf{C}_{i_{p+1}} & \mathbf{D}_{i_{p+1}i_1} & \mathbf{D}_{i_{p+1}i_2} & \cdots & \mathbf{D}_{i_{p+1}i_p} \\ \mathbf{C}_{i_{p+2}} & \mathbf{D}_{i_{p+2}i_1} & \mathbf{D}_{i_{p+2}i_2} & \cdots & \mathbf{D}_{i_{p+2}i_p} \\ \vdots & \vdots & \vdots & \ddots & \vdots \\ \mathbf{C}_{i_\nu} & \mathbf{D}_{i_\nu i_1} & \mathbf{D}_{i_\nu i_2} & \cdots & \mathbf{D}_{i_\nu i_p} \end{bmatrix} \tag{7.10}$$

In addition, since it is assumed that the mode σ_i is controllable and observable, the rank of the matrix (7.10) is equal to n for $p=0$ and $p=\nu$. Therefore, the condition $0 \leq p \leq \nu$ given above can be reduced to $1 \leq p \leq \nu-1$. It is clear that the rank of the matrix $\mathbf{A}-\sigma_i I_n$ is $n-1$, and also, the ith column and the ith row of this matrix are both zeros. Hence, if there exists a nonzero entry either in the ith column or in the ith row of the matrix given in (7.10), its rank will be at least n. As a result, the rank of the matrix in (7.10) is less than n, if and only if both of the following conditions hold:

i) All of the entries of the ith column and the ith row of the matrix given in (7.10) are zero, i.e., $\mathbf{b}_\eta^{i,\alpha} = \mathbf{c}_\gamma^{\beta,i} = 0$ for any α, β, η, and γ satisfying (7.9).

ii) The rank of the following matrix:

$$\begin{bmatrix} \sigma_1^i & \cdots & 0 & 0 & \cdots & 0 & \mathbf{b}_\eta^{1,\alpha} \\ \vdots & \ddots & \vdots & \vdots & \ddots & \vdots & \vdots \\ 0 & \cdots & \sigma_{i-1}^i & 0 & \cdots & 0 & \mathbf{b}_\eta^{i-1,\alpha} \\ 0 & \cdots & 0 & \sigma_{i+1}^i & \cdots & 0 & \mathbf{b}_\eta^{i+1,\alpha} \\ \vdots & \ddots & \vdots & \vdots & \ddots & \vdots & \vdots \\ 0 & \cdots & 0 & 0 & \cdots & \sigma_n^i & \mathbf{b}_\eta^{n,\alpha} \\ \mathbf{c}_\gamma^{\beta,1} & \cdots & \mathbf{c}_\gamma^{\beta,i-1} & \mathbf{c}_\gamma^{\beta,i+1} & \cdots & \mathbf{c}_\gamma^{\beta,n} & \mathbf{d}_{\gamma\eta}^{\beta,\alpha} \end{bmatrix} \tag{7.11}$$

(which is a sub-matrix of the one given by (7.10)) is less than n for any α, β, η, and γ satisfying (7.9), where $\sigma_j^i := \sigma_j - \sigma_i,\ i,j \in \{1,2,...,n\}$. Partition the matrix given by (7.11) into four sub-matrices, and denote it with $\begin{bmatrix} A_i & \Phi_1 \\ \Phi_2 & \mathbf{d}_{\gamma\eta}^{\beta,\alpha} \end{bmatrix}$, where $A_i \in \Re^{(n-1)\times(n-1)}$, $\Phi_1 \in \Re^{(n-1)\times 1}$, and $\Phi_2 \in \Re^{1\times(n-1)}$. Since the matrix A_i is nonsingular (because it is assumed that $\sigma_1,...,\sigma_n$ are distinct), one can write:

$$\det \begin{bmatrix} A_i & \Phi_1 \\ \Phi_2 & \mathbf{d}_{\gamma\eta}^{\beta,\alpha} \end{bmatrix} = \det(A_i) \times \det\left(\mathbf{d}_{\gamma\eta}^{\beta,\alpha} - \Phi_2 A_i^{-1}\Phi_1\right) \tag{7.12}$$

Thus, the rank of the matrix given in (7.11) is less than n, if and only if the scalar $\Phi_2 A_i^{-1}\Phi_1$ is equal to $\mathbf{d}_{\gamma\eta}^{\beta,\alpha}$, i.e.:

$$\sum_{\mu=1,\ \mu\neq i}^{n} \frac{\mathbf{b}_\eta^{\mu,\alpha}\mathbf{c}_\gamma^{\beta,\mu}}{\sigma_\mu^i} = \mathbf{d}_{\gamma\eta}^{\beta,\alpha} \tag{7.13}$$

■

Define now the matrix M_i as:

$$M_i := \mathbf{C} \times \mathrm{diag}\left(\left[\frac{1}{\sigma_1 - \sigma_i}, \ldots, \frac{1}{\sigma_{i-1} - \sigma_i}, 0, \frac{1}{\sigma_{i+1} - \sigma_i}, \ldots, \frac{1}{\sigma_\nu - \sigma_i}\right]\right)\mathbf{B} - \mathbf{D} \tag{7.14}$$

and denote its (μ_1, μ_2) block entry with $M_i^{\mu_1,\mu_2} \in \Re^{r_{\mu_1} \times m_{\mu_2}}$, for any $\mu_1, \mu_2 \in \bar{\nu}$. Note that the expression of M_i resembles that of the transfer function matrix of the system $\mathscr{S}$, while the sign of $\mathbf{D}$ is different in M_i.

Theorem 2 *The mode σ_i, $i \in \{1,2,...,n\}$, is a DFM of the system $\mathscr{S}$, $\nu \geq 2$, if and only if any of the following conditions holds:*

(i) The ith row of the matrices $\mathbf{B}_1, \mathbf{B}_2, ..., \mathbf{B}_\nu$ are zero.

ii) The ith column of the matrices $\mathbf{C}_1, \mathbf{C}_2, ..., \mathbf{C}_\nu$ are zero.

iii) There exist a permutation of $\{1,2,...,\nu\}$ denoted by distinct integers $i_1, i_2, ..., i_\nu$ and an integer $p \in [1, \nu-1]$ such that $M_i^{\gamma,\eta}$ is a zero matrix for any $\eta \in \{i_1, i_2, ..., i_p\}$ and $\gamma \in \{i_{p+1}, i_{p+2}, ..., i_\nu\}$, and moreover the ith row of the matrices $\mathbf{B}_1, \mathbf{B}_2, ..., \mathbf{B}_{i_p}$ and the ith column of $\mathbf{C}_{i_{p+1}}, \mathbf{C}_{i_{p+2}}, ..., \mathbf{C}_{i_\nu}$ are all zero.

Proof Criteria (i) and (ii) are equivalent to the uncontrollability and the unobservability, respectively. Furthermore, Criterion (iii) is resulted from Theorem 1, on noting that $M_i^{\gamma,\eta}$ is a $r_\gamma \times m_\eta$ matrix whose (β, α) entry is equal to:

$$\sum_{\mu=1,\ \mu\neq i}^{n} \frac{\mathbf{b}_\eta^{\mu,\alpha}\mathbf{c}_\gamma^{\beta,\mu}}{\sigma_\mu - \sigma_i} - \mathbf{d}_{\gamma\eta}^{\beta,\alpha} \tag{7.15}$$

for any $\beta \in [1, r_\gamma]$, $\alpha \in [1, m_\eta]$. ■

It is desired now to construct a graph based on the matrix M_i. Consider a bipartite graph $\mathscr{G}_i$ with ν vertices $1, 2, ..., \nu$ in each of its vertex sets, namely set 1 and set 2. For any $\mu_1, \mu_2 \in \bar{\nu}$, connect vertex μ_1 of set 1 to vertex μ_2 of set 2 if the matrix $M_i^{\mu_1,\mu_2}$ is a zero matrix. Then, mark vertex μ_1 of set 1 if the ith column of the matrix C_{μ_1} is a zero vector, for any $\mu_1 \in \bar{\nu}$. Likewise, mark vertex μ_2 of set 2 if the ith row of the matrix B_{μ_2} is a zero vector, for any $\mu_2 \in \bar{\nu}$.

The following algorithm results from Theorem 2 for verifying whether or not the mode σ_i is a DFM of the system $\mathscr{S}$.

Algorithm 1

Step 1) *Compute the matrix* M_i*, and construct the graph* $\mathscr{G}_i$ *in terms of it, as pointed out earlier.*

Step 2) *Verify if all of the vertices in set 1 of the graph* $\mathscr{G}_i$ *are marked. If yes, go to Step 6.*

Step 3) *Verify if all of the vertices in set 2 of the graph* $\mathscr{G}_i$ *are marked. If yes, go to Step 6.*

Step 4) *Check whether the graph* $\mathscr{G}_i$ *includes a complete bipartite subgraph such that all of its vertices are marked and moreover the set of the indices of its vertices is equal to the set* $\bar{\nu}$*. If yes, go to Step 6.*

Step 5) *The mode* σ_i *is not a DFM of the system* $\mathscr{S}$*. Stop the algorithm.*

Step 6) *The mode* σ_i *is a DFM of the system* $\mathscr{S}$*. Stop the algorithm.*

Algorithm 1 proposes a simple graph-theoretic approach to find the DFMs of the system $\mathscr{S}$. This method requires deriving a certain matrix, and then checking the existence of a complete subgraph in a graph, which can be accomplished using numerous efficient algorithms. In contrast, the existing methods require the rank of several matrices (say 2^ν) to be checked, which can be cumbersome when the matrix is of high dimension. In fact, the above algorithm presents a simple combinatorial procedure as a more efficient alternative to find the DFMs of a system (with distinct modes).

Corollary 1 *Denote the number of matrices* $\mathbf{B}_1, \mathbf{B}_2, ..., \mathbf{B}_\nu$ *whose* i^{th} *row are zero with* Γ_i*. Furthermore, denote the number of matrices* $\mathbf{C}_1, \mathbf{C}_2, ..., \mathbf{C}_\nu$ *whose* i^{th} *column are zero with* $\bar{\Gamma}_i$*. If* $\Gamma_i + \bar{\Gamma}_i$ *is less than* ν*, then* σ_i *is not a DFM of the system* $\mathscr{S}$*.*

Proof It is straightforward to show that if $\Gamma_i + \bar{\Gamma}_i$ is less than ν, none of Steps 1, 2 or 3 of Algorithm 1 is fulfilled. ■

Corollary 1 presents a quite simple test as a sufficient condition to verify whether σ_i can be a DFM of the system or not.

7.5 Characterization of quotient fixed modes

It is desired now to present a graph-theoretic approach to obtain the QFMs of the system $\mathscr{S}$, similar to the one introduced for the DFMs in the preceding section. Since QFM is merely defined for the strictly proper systems, it will be assumed hereafter that $D = \mathbf{D} = 0$.

Theorem 3 *The mode σ_i is a QFM of the system $\mathscr{S}$, $\nu \geq 2$ if and only if either condition (a) or condition (b) given below holds:*

a) σ_i is an uncontrollable or unobservable mode.

b) There exist a permutation of $\{1, 2, ..., \nu\}$ denoted by distinct integers $i_1, i_2, ..., i_\nu$ and an integer $p \in [1, \nu - 1]$ such that for all η and γ given by:

$$\eta \in \{i_1, i_2, ..., i_p\}, \quad \gamma \in \{i_{p+1}, i_{p+2}, ..., i_\nu\} \tag{7.16}$$

both of the conditions given below hold:

i) The ith column of the matrix $\mathbf{C}_\gamma$ and the ith row of the matrix $\mathbf{B}_\eta$ are both zero vectors.

ii) Consider the jth column of the matrix $\mathbf{C}_\gamma$ and the jth row of the matrix $\mathbf{B}_\eta$; at least one of these two vectors is zero, for any $j \in \{1, 2, ..., n\}$.

Proof It is trivial to show that if condition (a) in Theorem 3 holds, the mode σ_i will be a QFM of the system $\mathscr{S}$. If it does not hold, then it follows directly from Theorems 1 and 2 given in [3], that σ_i is a QFM if and only if there exist a permutation of $\{1, 2, ..., \nu\}$ denoted by distinct integers $i_1, i_2, ..., i_\nu$ and an integer $p \in [1, \nu - 1]$ such that $\mathbf{b}_\eta^{i,\alpha} = \mathbf{c}_\gamma^{\beta,i} = 0$, and $\mathbf{b}_\eta^{\mu,\alpha}\mathbf{c}_\gamma^{\beta,\mu}$=0 for all η, γ, α and β given by (7.9) and $\mu \in \{1, 2..., n\}$. It is straightforward to show that this requirement is identical to condition (b) in the theorem. ■

Consider a bipartite graph $\bar{\mathscr{G}}_i$ with ν vertices $1, 2, ..., \nu$ in each of its vertex sets, namely set 1 and set 2. For any $\mu_1, \mu_2 \in \bar{\nu}$, connect vertex μ_1 of set 1 to vertex μ_2 of set 2 if either the jth column of $\mathbf{C}_{\mu_1}$ or the jth row of $\mathbf{B}_{\mu_2}$ is a zero vector for all $j \in \{1, 2, ..., n\}$. Then, mark vertex μ_1 of set 1 if the ith column of the matrix $\mathbf{C}_{\mu_1}$ is a zero vector, for any $\mu_1 \in \bar{\nu}$. Likewise, mark vertex μ_2 of set 2 if the ith row of the matrix $\mathbf{B}_{\mu_2}$ is a zero vector, for any $\mu_2 \in \bar{\nu}$. It is worth noting that the graphs $\bar{\mathscr{G}}_1, \bar{\mathscr{G}}_2, ..., \bar{\mathscr{G}}_\nu$ have the same edges, although the marking of their vertices might be different.

The following algorithm results from Theorem 3 for verifying whether or not the mode σ_i is a QFM of the system $\mathscr{S}$.

Algorithm 2

Step 1) *Construct the graph* $\bar{\mathscr{G}}_i$*, as discussed above.*

Step 2) *Verify if all of the vertices in set 1 of the graph* $\mathscr{G}_i$ *are marked. If yes, go to Step 6.*

Step 3) *Verify if all of the vertices in set 2 of the graph* $\mathscr{G}_i$ *are marked. If yes, go to Step 6.*

Step 4) *Check whether the graph* $\bar{\mathscr{G}}_i$ *includes a complete bipartite subgraph such that all of its vertices are marked and the set of the indices of its vertices is equal to the set* $\bar{\mathrm{v}}$*. If yes, go to Step 6.*

Step 5) *The mode* σ_i *is not a QFM of the system* $\mathscr{S}$*. Stop the algorithm.*

Step 6) *The mode* σ_i *is a QFM of the system* $\mathscr{S}$*. Stop the algorithm.*

7.6 Illustrative examples

Example 1 Consider a system $\mathscr{S}$ consisting of five single-input single-output (SISO) subsystems with the following state-space matrices:

$$\mathbf{A}=\begin{bmatrix}1&0&0&0&0\\0&2&0&0&0\\0&0&3&0&0\\0&0&0&4&0\\0&0&0&0&5\end{bmatrix},\quad \mathbf{B}=\begin{bmatrix}0&0&0&1&2\\2&1&3&1&4\\0&2&4&-1&5\\0&0&3&0&-3\\0&0&0&3&-1\end{bmatrix},$$
$$\mathbf{C}=\begin{bmatrix}0&3&2&1&4\\0&3&4&2&-1\\5&4&3&-2&4\\0&2&3&1&3\\0&-2&-3&-2&-4\end{bmatrix},\quad \mathbf{D}=\begin{bmatrix}6&5&14&3&2\\6&7&19&4&2\\8&7&16&-2&-4\\4&5&13&0&1\\-4&-5&-14&-1&2\end{bmatrix} \tag{7.17}$$

It is desired to verify which of the modes $\sigma_i = i,\ i \in \bar{\nu} = \{1,2,3,4,5\}$, are DFMs of the system $\mathscr{S}$. First, let the test given in Corollary 1 be carried out. Since the first entries of $\mathbf{B}_1, \mathbf{B}_2, \mathbf{B}_3, \mathbf{C}_1, \mathbf{C}_2, \mathbf{C}_4$ and $\mathbf{C}_5$ are all zero, $\Gamma_1 + \bar{\Gamma}_1$ is equal to 7. Similarly, one can conclude that:

$$\Gamma_2 + \bar{\Gamma}_2 = 0,\ \ \Gamma_3 + \bar{\Gamma}_3 = 1,\ \ \Gamma_4 + \bar{\Gamma}_4 = 3,\ \ \Gamma_5 + \bar{\Gamma}_5 = 3 \tag{7.18}$$

Due to the fact that $\Gamma_i + \bar{\Gamma}_i < 5$ for $i = 2,3,4,5$, it follows from Corollary 1 that none of the modes $2,3,4$ and 5 is a DFM of the system $\mathscr{S}$. Algorithm 1 will now be used to find out whether $\sigma_1 = 1$ is a DFM. The matrix M_1 will be obtained as:

$$M_1 = \mathbf{C} \times \text{diag}\left(\left[0\ ,\ 1\ ,\ \frac{1}{2}\ ,\ \frac{1}{3}\ ,\ \frac{1}{4}\right]\right)\mathbf{B} - \mathbf{D} = \begin{bmatrix} 0 & 0 & 0 & 2 & 13 \\ 0 & 0 & 0 & -3.75 & 18.25 \\ 0 & 0 & 0 & 7.5 & 28.5 \\ 0 & 0 & 0 & 2.75 & 12.75 \\ 0 & 0 & 0 & -2.5 & -14.5 \end{bmatrix} \tag{7.19}$$

The graph $\mathscr{G}_1$ corresponding to the matrix M_1 is sketched in Figure 7.3. Since the first entries of $\mathbf{B}_1, \mathbf{B}_2, \mathbf{B}_3, \mathbf{C}_1, \mathbf{C}_2, \mathbf{C}_4$ and $\mathbf{C}_5$ are all zero, vertices 1, 2 and 3 from set 2, and vertices 1, 2, 4 and 5 from set 1 of the graph $\mathscr{G}_1$ are marked by filled circles, as shown in the figure. It can be easily observed that vertices 4, 5 of set 1 and vertices 1, 2, 3 of set 2 fulfill the following criteria:

- All of them are marked.
- They constitute a complete bipartite graph.
- The set of their labels is equal to $\bar{\nu}$.

Therefore, $\sigma_1 = 1$ is a DFM of the system (from Step 4 of Algorithm 2).

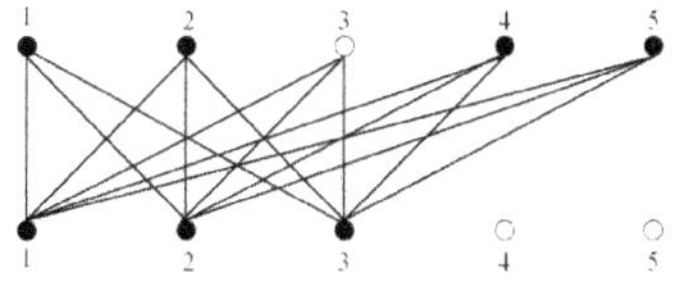

Figure 7.3: The graph $\mathscr{G}_1$ corresponding to the matrix M_1 given in (7.19).

Regarding the mode $\sigma_3 = 3$, let Algorithm 2 be pursued for this mode regardless of the observation that it failed the test given in Corollary 1. The matrix M_3 is equal to:

$$M_3 = \mathbf{C} \times \text{diag}\left(\left[-1\,,\,0\,,\,1\,,\,\frac{1}{2}\,,\,\frac{1}{3}\right]\right)\mathbf{B} - \mathbf{D} = \begin{bmatrix} -12 & -8 & -20 & 0 & -19 \\ -12 & -10 & -22 & -8.5 & -19 \\ -16 & -11 & -34 & 1.5 & -13 \\ -8 & -7 & -16 & 2.5 & -13 \\ 8 & 7 & 14 & -3 & 14 \end{bmatrix} \quad (7.20)$$

The corresponding graph $\mathscr{G}_3$ is depicted in Figure 7.4. Since there are not enough edges in the graph to create a complete bipartite subgraph which spans all the indices, thus $\sigma_3 = 3$ is not a DFM of the system (which also confirms the result obtained from Corollary 1).

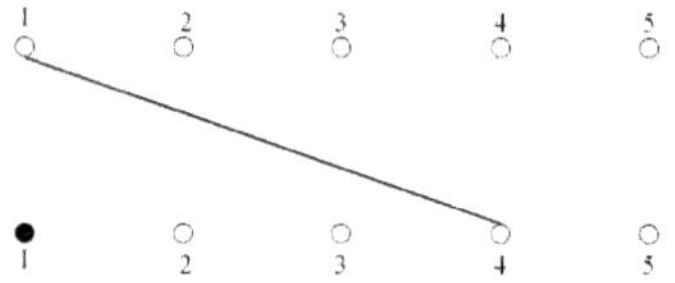

Figure 7.4: The graph $\mathscr{G}_3$ corresponding to the matrix M_3 given in (7.20).

Consequently, the system has only one DFM at 1. This result could also be obtained by using the method given in [2] or [15], which require the rank of 5×2^5 matrices with the dimensions between 5 and 10 be checked. The sizable difference between the computational requirements of the method presented in this chapter and the ones given in [2; 15] demonstrates the efficacy of this work. It is worth mentioning that the results obtained here by using the proposed method are attained by hand, while the methods given in [2; 15] require a proper software (such as MATLAB).

Example 2 Consider a strictly proper system $\mathscr{S}$ consisting of three two-input two-output (SISO) subsystems with the following parameters:

$$
\mathbf{A}=\begin{bmatrix}1&0&0&0&0\\0&2&0&0&0\\0&0&3&0&0\\0&0&0&4&0\\0&0&0&0&5\end{bmatrix},\ \mathbf{B}_1=\begin{bmatrix}0&0\\0&0\\0&0\\2&-1\\0&0\end{bmatrix},\ \mathbf{B}_2=\begin{bmatrix}0&0\\2&1\\0&0\\1&2\\0&0\end{bmatrix},\ \mathbf{B}_3=\begin{bmatrix}1&1\\-1&-1\\1&-1\\-1&1\\-1&-1\end{bmatrix},
$$

$$
\mathbf{C}_1=\begin{bmatrix}1&-1\\-1&1\\1&-1\\-1&1\\1&-1\end{bmatrix}^T,\ \mathbf{C}_2=\begin{bmatrix}2&3\\0&0\\0&0\\0&0\\1&1\end{bmatrix}^T,\ \mathbf{C}_3=\begin{bmatrix}5&6\\0&0\\0&0\\0&0\\-1&-1\end{bmatrix}^T \tag{7.21}
$$

The graphs $\bar{\mathscr{G}}_i$, $i\in\{1,2,3,4,5\}$ and $\bar{\mathscr{G}}_3$ are depicted in Figures 7.5, 7.6 and 7.7. Using Algorithm 2, it can be concluded that $\lambda=2$ is a QFM of the system, as step 4 will be satisfied by considering vertices 2 and 3 from set 1, and vertex 1 from set 2. Likewise, the mode $\lambda=3$ is a QFM by considering either *vertices 2 and 3 from set 1, and vertex 1 from set 2* or *vertex 3 from set 1, and vertices 1 and 2 from set 2*. It can be easily verified that none of the remaining modes are QFMs of the system $\mathscr{S}$.

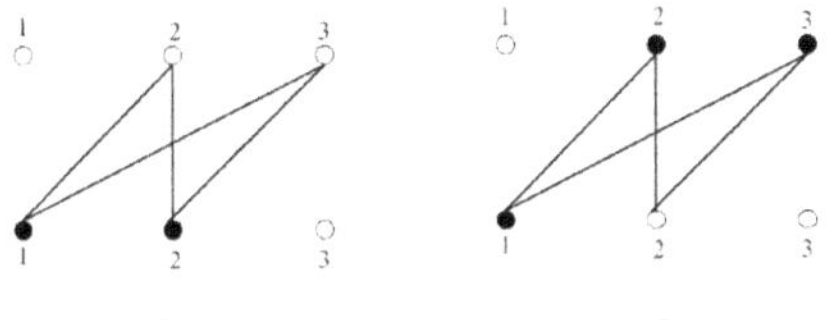

Figure 7.5: The graphs $\bar{\mathscr{G}}_1$ and $\bar{\mathscr{G}}_2$ are sketched in (a) and (b), respectively.

7.7 Conclusions

This chapter aims to characterize the fixed modes of a decentralized system with distinct modes. First, decentralized fixed modes (DFM) are described using graph-theoretic techniques. Then, quotient fixed modes (QFM), which are immovable with respect to any type of decentralized control law, are

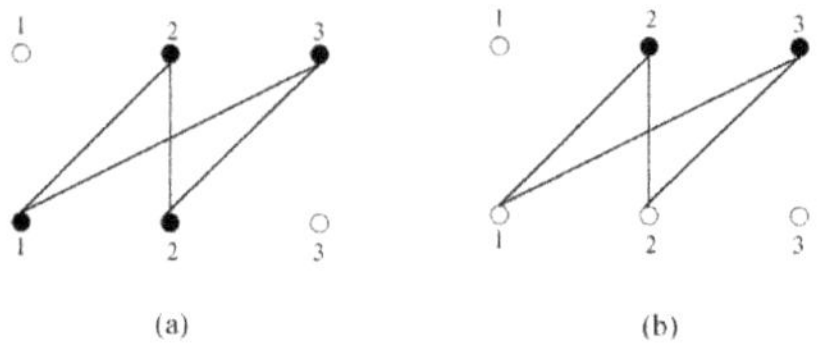

Figure 7.6: The graphs $\bar{\mathscr{G}}_3$ and $\bar{\mathscr{G}}_4$ are sketched in (a) and (b), respectively.

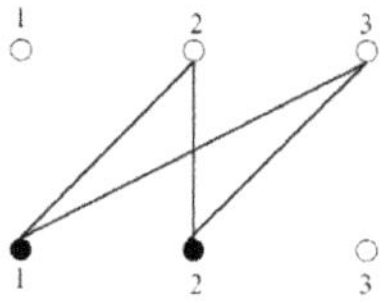

Figure 7.7: The graphs $\bar{\mathscr{G}}_5$.

characterized. Unlike the existing methods which require the computation of the rank of several matrices, the approaches proposed here transform the knowledge of the system into bipartite graphs. Then, it is asserted that finding a complete bipartite subgraph with a certain property is equivalent to the existence of a DFM. A similar result is attained for the QFMs. The efficacy of the proposed method is demonstrated through numerical examples.

Bibliography

[1] S. H. Wang and E. J. Davison, "On the stabilization of decentralized control systems," *IEEE Transactions on Automatic Control*, vol. 18, no. 5, pp. 473-478, 1973.

[2] E. J. Davison and T. N. Chang, "Decentralized stabilization and pole assignment for general proper systems," *IEEE Transactions on Automatic Control*, vol. 35, no. 6, pp. 652-664, 1990.

[3] J. Lavaei and A. G. Aghdam, "Elimination of fixed modes by means of high-performance constrained periodic control," *in Proc. of 45th IEEE Conference on Decision and Control*, San Diego, CA, 2006.

[4] D. D. Šiljak, Decentralized control of complex systems, Cambridge: Academic Press, 1991.

[5] D. D. Šiljak and A. I. Zecevic, "Control of large-scale systems: Beyond decentralized feedback," *Annual Reviews in Control*, vol. 29, no. 2, pp. 169-179, 2005.

[6] J. Lavaei and A. G. Aghdam, "A necessary and sufficient condition for the existence of a LTI stabilizing decentralized overlapping controller," *in Proc. of 45th IEEE Conference on Decision and Control*, San Diego, CA, 2006.

[7] G. Inalhan, D. M. Stipanovic, and C. J. Tomlin, "Decentralized optimization with application to multiple aircraft coordination," *Proceedings of the 41st IEEE Conference on Decision and Control*, Vegas, NV., pp. 1147-1155, 2002.

[8] J. Lavaei, A. Momeni and A. G. Aghdam, "High-performance decentralized control for formation flying with leader-follower structure," *in Proc. of 45th IEEE Conference on Decision and Control*, San Diego, CA, 2006.

[9] J. Lavaei and A. G. Aghdam, "Decentralized control design for interconnected systems based on a centralized reference controller," *in Proc. of 45th IEEE Conference on Decision and Control*, San Diego, CA, 2006.

[10] S. S. Stankovic, M. J. Stanojevic, and D. D. Šiljak, "Decentralized overlapping control of a platoon of vehicles," *IEEE Transactions on Control Systems Technology*, vol. 8, no. 5, pp. 816-832, 2000.

[11] J. Lavaei and A. G. Aghdam, "Optimal periodic feedback design for continuous-time LTI systems with constrained control structure," *International Journal of Control*, vol. 80, no. 2, pp. 220-230, Feb. 2007.

[12] J. Lavaei and A. G. Aghdam, "Simultaneous LQ control of a set of LTI systems using constrained generalized sampled-data hold functions," *Automatica*, vol. 43, no. 2, pp. 274-280, Feb. 2007.

[13] B. L. O. Anderson and D. J. Clements, "Algebraic characterizations of fixed modes in decentralized systems," *Automatica*, vol. 17, no. 5, pp. 703-712, 1981.

[14] B. L. O. Anderson, "Transfer function matrix description of decentralized fixed modes," *IEEE Transactions on Automatic Control*, wol. 27, no. 6, pp. 1176-1182, 1982.

[15] E. J. Davison and S. H. Wang, "A characterization of decentralized fixed modes in terms of transmission zeros," *IEEE Transactions on Automatic Control*, vol. 30, no. 1, pp. 81-82, 1985.

[16] Z. Gong and M. Aldeen, "On the characterization of fixed modes in decentralized control," *IEEE Transactions on Automatic Control*, vol. 37, no. 7, pp. 1046-1050, 1992.

[17] Z. Gong and M. Aldeen, "Stabilization of decentralized control systems," *Journal of Mathematical Systems, Estimation, and Control*, vol. 7, no. 1, pp. 1-16, 1997.

Chapter 8

Decentralized Overlapping Control: Stabilizability and Pole-Placement

8.1 Abstract

This chapter deals with the control of the large-scale interconnected systems with a constrained control structure. It is shown that ceratin modes of the system can be freely placed anywhere on the complex plane, by using a linear time-invariant (LTI) structurally constrained controller. These modes have been identified by introducing the notion of a decentralized overlapping fixed mode (DOFM). This implies that the system is stabilizable by a LTI structurally constrained controller, if and only if it does not have any unstable DOFM. Furthermore, a design procedure is proposed for obtaining a stabilizing controller to achieve the desired pole placement for the systems with no DOFM. In addition, the problem of designing a structurally constrained optimal LTI controller with respect to a quadratic performance index is studied. Designing various types of structurally constrained controllers, such as periodic feedback, is then investigated. The notion of a quotient overlapping fixed mode (QOFM) is also introduced, and it is shown that a system is stabilizable by mean of a general controller, i.e. nonlinear and time-varying, if and only if it does not have any unstable QOFM . In the case of no unstable QOFM, it is proved that there exists a finite-dimensional linear time-varying structurally constrained controller to stabilize the system.

8.2 Introduction

In the past three decades, the problem of decentralized control has been thoroughly investigated in the literature, and a variety of its aspects are studied [1; 2; 3]. More recently, the problem of decentralized overlapping control has attracted several researchers [4; 5]. The decentralized overlapping control is fundamentally used in two cases:

i) when the subsystems of a system (referred to as overlapping subsystems) share some states [6; 7; 8]. In this case, it is usually desired that the structure of the controller matches the overlapping structure of the system [8];

ii) when there are some limitations on the availability of the states. In this case, only certain outputs of the system are available for constructing each control signal.

The control constraint in both cases discussed above can be represented by a binary information flow matrix. For instance, when this matrix is block diagonal with the entries of the main diagonal blocks all equal to 1, the control structure is decentralized, and when all of its entries are 1, the controller is centralized. One particular structural constraint for the controller, which is investigated intensively in the literature, corresponds to an information flow matrix whose entries on the main diagonal blocks, as well as the last block column and the last block row are all equal to 1. This is often referred to as bordered block-diagonal structure (BBD) or block array structure (BAS), and has found several practical applications [8; 9; 10]. In general, for an interconnected system with a given information flow matrix, the following open questions are of main interest in the literature:

1. Does there exist a stabilizing static output feedback controller for the system?

2. Does there exist a linear time-invariant (LTI) controller to stabilize the system, if there is no static one?

3. How can a static or dynamic LTI controller be found such that a predefined quadratic performance index is minimized?

4. Can the poles of the system be placed at any arbitrary locations, when there exists a LTI stabilizing controller for the system?

5. Can the system be stabilized by a non-LTI controller when a LTI stabilizing controller does not exist?

The first three questions have been addressed in the literature in the decentralized overlapping control framework. This is accomplished by using a transformation which expands the structure of the system such that the resultant control configuration is decentralized. Then, by using the existing design techniques, the desired decentralized controller is obtained for the expanded system. The last step of the design is to contract the controller obtained in order to make it suitable for the original system. This approach is substantially useful, when the structure of the system itself is overlapping as well because in that case, the subsystems of the expanded system are disjoint [11]. Nevertheless, one of the shortcomings of this method is that the expanded system is inherently uncontrollable, and thus, this design approach may not be useful in general. This problem has been addressed in several papers, e.g. see [8], [4]. Furthermore, the contraction of the designed controller can cause some problems in general. Although a large number of conditions for contraction are presented, finding a proper contraction is still an open problem [8]. In addition, it is often assumed that a static state feedback controller (as opposed to a general output feedback controller) is to be designed, which may not be suitable in practical applications. In the special case of a BAS control design, a number of methods have been proposed in the literature, including an optimal control design technique [9; 10]. The other existing methods for designing a BAS or an overlapping (or structurally constrained) controller often present some sufficient conditions in the form of LMI, and fail to address some of the important questions discussed above [8; 12; 13]. Furthermore, these methods assume that the system is strictly proper, while the generalization of the methods to general proper systems is not straightforward. The present work addresses the problem of designing a structurally constrained controller, and is aimed to answer the open questions discussed above, for any LTI system with any arbitrary information flow structure.

8.3 Problem formulation

Consider a LTI interconnected system $\mathscr{S}$ consisting of ν subsystems with the following state-space representation:

$$\begin{aligned}\dot{x}(t) &= Ax(t) + \sum_{i=1}^{\nu} B_i u_i(t) \\ y_i(t) &= C_i x(t) + \sum_{j=1}^{\nu} D_{ij} u_j(t), \quad i \in \bar{\nu} := \{1,2,...,\nu\}\end{aligned} \tag{8.1}$$

where $x(t) \in \Re^n$ is the state, and $u_i(t) \in \Re^{m_i}$ and $y_i(t) \in \Re^{r_i}$, $i \in \bar{\nu}$, are the input and the output of the ith subsystem S_i, respectively. Define the following matrices:

$$B := \begin{bmatrix} B_1 & B_2 & \cdots & B_\nu \end{bmatrix}, \quad C := \begin{bmatrix} C_1 \\ C_2 \\ \vdots \\ C_\nu \end{bmatrix}, \quad D := \begin{bmatrix} D_{11} & \cdots & D_{1\nu} \\ \vdots & \ddots & \vdots \\ D_{\nu 1} & \cdots & D_{\nu\nu} \end{bmatrix} \tag{8.2}$$

Define also:

$$m := \sum_{i=1}^{\nu} m_i, \quad r := \sum_{i=1}^{\nu} r_i \tag{8.3}$$

It is desired to stabilize the system $\mathscr{S}$ by using a structurally constrained controller. These constraints determine which outputs y_j $(j \in \bar{\nu})$ are available to construct any specific input u_i $(i \in \bar{\nu})$ of the system. In order to simplify the formulation of the control constraint, a matrix $\mathscr{K}$ with binary elements is defined, where its (i,j) block entry, $i,j \in \bar{\nu}$, is a $m_i \times r_j$ matrix whose elements are all equal to 1 if the output y_j can contribute to the construction of the input u_i, and is a $m_i \times r_j$ zero matrix otherwise. The matrix $\mathscr{K}$ represents the control constraint, and will be referred to as the information flow matrix.

To represent the structural constraint of the system, the corresponding information flow matrix is enclosed in parentheses throughout the chapter, if necessary. For instance, $\mathscr{S}(\mathscr{K})$ indicates that the structure of the controller to be designed for the system $\mathscr{S}$ is to comply with the information flow matrix $\mathscr{K}$.

In the special case, when the entries of the matrix $\mathscr{K}$ are all equal to 1, the corresponding controller is centralized, and when $\mathscr{K}$ is block diagonal, the corresponding controller is decentralized. Throughout this chapter, the term "decentralized controller" is referred to the set of local controllers for an interconnected system with a block diagonal information flow matrix.

It is to be noted that in the case of a block diagonal matrix $\mathscr{K}$, one can use the existing methods, e.g. [2], to find the decentralized fixed modes (DFM) of the system, if any. Then, if the system does not have any DFM in the closed right-half plane (RHP), one can use a LTI decentralized controller to stabilize it and place those modes which are not fixed, in any arbitrary location in the complex plane. Furthermore, the system can still be stabilized in the presence of unstable DFMs, as long as they are not quotient fixed modes (QFM) [3]. A system with unstable QFMs cannot be stabilized by using any type of controller, i.e., nonlinear and time-varying. However, there is no necessary and sufficient condition for the existence of a general stabilizing controller, when $\mathscr{K}$ is not block diagonal. This problem will be addressed in the following sections. It is to be noted that to avoid trivial cases (i.e., standard decentralized and centralized systems), the matrix $\mathscr{K}$ will hereafter be assumed not to be block diagonal, and to have at least one zero block.

8.4 Computing the transformation matrices

Definition 1 *Consider two arbitrary systems $\mathscr{S}_{d_1}$ and $\mathscr{S}_{d_2}$ associated with the information flow matrices $\mathscr{K}_{d_1}$ and $\mathscr{K}_{d_2}$, where $\mathscr{S}_{d_1}$ and $\mathscr{S}_{d_2}$ are of the same order and have the same initial state. Let* $\mathbf{M}$ *denote a given set of controllers. The systems $\mathscr{S}_{d_1}(\mathscr{K}_{d_1})$ and $\mathscr{S}_{d_2}(\mathscr{K}_{d_2})$ are called* analogous *with respect to* $\mathbf{M}$ *if for any controller K_{d_1} in* $\mathbf{M}$ *complying with the information flow matrix $\mathscr{K}_{d_1}$, there also exists a controller K_{d_2} in* $\mathbf{M}$ *complying with the information flow matrix $\mathscr{K}_{d_2}$ (and vice versa), such that the state of the system $\mathscr{S}_{d_1}$ under the controller K_{d_1} is equivalent to the state of $\mathscr{S}_{d_2}$ under K_{d_2}.*

The motivation for introducing the notion of analogous systems is that given a system $\mathscr{S}(\mathscr{K})$ with any general information flow structure $\mathscr{K}$, it is desired to find an analogous system with a decentralized (i.e. block diagonal) information flow structure. It is to be noted there are several efficient methods for design of decentralized controllers. Thus, the problem reduces to designing a proper decentralized controller, and finding a transformation to change the block-diagonal structure of the controller to the desired structure for the original system $\mathscr{S}(\mathscr{K})$. This is an indirect method of design, which unlike the existing indirect approaches aims to identify the fixed modes with respect to a structurally constrained controller. This section presents some transformation matrices which will later be used to construct systems *analogous* to the system $\mathscr{S}(\mathscr{K})$.

Define the control *interaction* structure **K** as a matrix whose (i, j) block entry, $i, j \in \bar{\nu}$, is a $m_i \times r_j$ matrix denote by k_{ij} if the output of the jth subsystem can contribute to the construction of the input of the ith subsystem, and is a $m_i \times r_j$ zero matrix otherwise. Note that k_{ij} represents a component of the controller, which transforms the output of the jth subsystem to the input of the ith subsystem. Note also that the interaction structure matrix **K** not only conveys the information of the matrix $\mathscr{K}$, but also labels the control components.

Procedure 1 *Construct the graph $\mathscr{G}$ as follows:*

1. *Define two sets of ν vertices. Label the sets as set 1 and set 2, and the vertices in each set as vertex 1 to vertex ν.*

2. *For any $i, j \in \bar{\nu}$, connect the ith vertex of set 1 to the jth vertex of set 2 with an edge, if the (i, j) block entry of $\mathscr{K}$ is not a zero matrix, i.e., if the output of the jth subsystem can contribute to the construction of the input of the ith subsystem. Label this edge with k_{ij}.*

As an example, consider a system consisting of four subsystems with the following control interaction structure matrix:

$$\mathbf{K} = \begin{bmatrix} k_{11} & 0 & 0 & 0 \\ k_{21} & k_{22} & 0 & k_{24} \\ k_{31} & 0 & k_{33} & 0 \\ 0 & k_{42} & 0 & k_{44} \end{bmatrix} \tag{8.4}$$

The graph $\mathscr{G}$ corresponding to the matrix **K** given above is depicted in Figure 8.1.

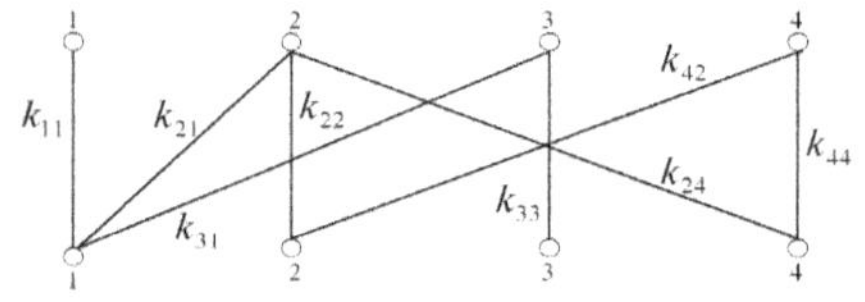

Figure 8.1: The graph $\mathscr{G}$ corresponding to the matrix **K** given by (8.4).

Procedure 2 *Partition the graph $\mathscr{G}$ into a set of complete bipartite subgraphs such that each edge of the graph $\mathscr{G}$ appears in only one of the subgraphs. It is to be noted that this partition may require some of the vertices of the graph $\mathscr{G}$ to appear in several subgraphs.*

It can be easily verified that Procedure 2 does not necessarily lead to a unique graph. Denote all the graphs which can be obtained through this procedure, with $\mathscr{G}_1, \mathscr{G}_2, ..., \mathscr{G}_l$. Without loss of generality, assume that $\mathscr{G}_1$ and $\mathscr{G}_l$ are the ones with the following properties:

- $\mathscr{G}_1$ is obtained by considering any vertex in set 1 of the graph $\mathscr{G}$ along with all of the vertices in set 2 connected to that vertex as a complete bipartite graph.
- $\mathscr{G}_l$ is obtained by considering any edge in the graph $\mathscr{G}$ as a complete bipartite graph.

As an example, consider again the graph $\mathscr{G}$ sketched in Figure 8.1. The graph $\mathscr{G}_2$ for this graph can be considered as the one depicted in Figure 8.2 (note that this graph is denoted by $\mathscr{G}_2$ instead of $\mathscr{G}_1$, because it does not satisfy the property of $\mathscr{G}_1$ described above). It is obvious from Figure 8.2 that, in this particular example, vertices 2 and 3 of the first set of vertices of $\mathscr{G}$ are repeated twice in $\mathscr{G}_2$.

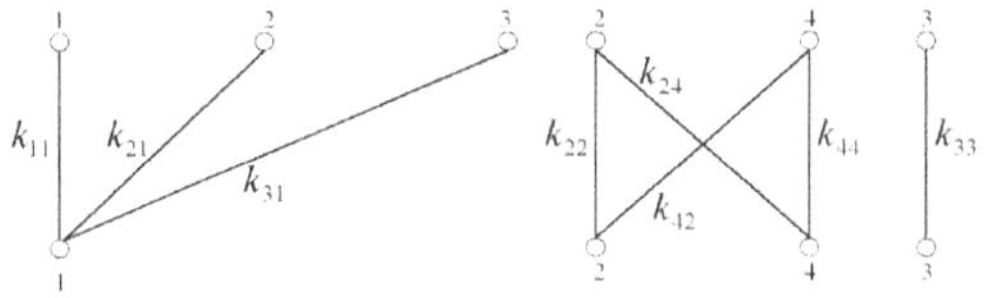

Figure 8.2: A decentralized graph $\mathscr{G}_2$ obtained from the graph $\mathscr{G}$ in Figure 8.1.

The following procedure can be used to construct the matrix $\mathbf{K}_\mu$ corresponding to the graph $\mathscr{G}_\mu$ for any $\mu \in \bar{l} := \{1, 2, ..., l\}$.

Procedure 3 *Label the complete bipartite subgraphs of $\mathscr{G}_\mu$ ($\mu \in \bar{l}$) as subgraphs 1 to ν_μ. Consider subgraph number σ ($\forall\, \sigma \in \{1, 2, ..., \nu_\mu\}$). Label those vertices of this subgraph which belong to set 1 as vertex $1, ..., \eta_\sigma^\mu$. This group of vertices will be referred to as subset 1 (corresponding to subgraph number σ). Similarly, label those vertices which belong to set 2 of this subgraph as vertex $1, ..., \bar{\eta}_\sigma^\mu$, and define subset 2 accordingly. Define $\mathbf{K}_\mu$ as a block diagonal matrix, where its (σ, σ) block entry, $\sigma = 1, ..., \nu_\mu$, is a matrix itself, whose (i, j) block entry is equal to the gain of the edge connecting vertex i of subset 1 to vertex j of subset 2 in subgraph number σ of $\mathscr{G}_\mu$, for any $i \in \{1, ..., \eta_\sigma^\mu\}$, $j \in \{1, ..., \bar{\eta}_\sigma^\mu\}$. Denote the dimension of the (σ, σ) block entry of $\mathbf{K}_\mu$ with $m_\sigma^\mu \times r_\sigma^\mu$, for $\sigma = 1, 2, ..., \nu_\mu$, and the dimension of $\mathbf{K}_\mu$ with $m^\mu \times r^\mu$.*

Using Procedure 3 and for a particular numbering of vertices in each subgraph of $\mathscr{G}_2$ in Figure 8.2, the following block diagonal matrix $\mathbf{K}_2$ is obtained:

$$\mathbf{K}_2 = \begin{pmatrix} k_{11} & 0 & 0 & 0 \\ k_{21} & 0 & 0 & 0 \\ k_{31} & 0 & 0 & 0 \\ 0 & k_{22} & k_{24} & 0 \\ 0 & k_{42} & k_{44} & 0 \\ 0 & 0 & 0 & k_{33} \end{pmatrix} \tag{8.5}$$

Remark 1 *It can be easily concluded from Procedures 1, 2 and 3, that there exists an onto mapping between the nonzero block entries of the matrix $\mathbf{K}_\mu$ and those of the matrix $\mathbf{K}$ for any $\mu \in \bar{l}$.*

Theorem 1 *There exist constant matrices Φ_μ and $\bar{\Phi}_\mu$ satisfying the following relation:*

$$\mathbf{K} = \Phi_\mu \mathbf{K}_\mu \bar{\Phi}_\mu \tag{8.6}$$

for any $\mu \in \bar{l}$.

Proof It is straightforward to show (by using Procedures 1, 2 and 3) that the matrix $\mathbf{K}_\mu$ can alternatively be constructed from $\mathbf{K}$ through a sequence of $L_\mu - 1$ operations (where L_μ is a finite number), such that the matrix $\mathbf{K}^\mu_{j+1}$ is formed in terms of $\mathbf{K}^\mu_j$ in the jth operation, for any $j \in \{1, 2, ..., L_\mu - 1\}$, where $\mathbf{K} = \mathbf{K}^\mu_1$ and $\mathbf{K}_\mu = \mathbf{K}^\mu_{L_\mu}$. Moreover, $\mathbf{K}^\mu_{j+1}$ is obtained from $\mathbf{K}^\mu_j$ for any $j \in \{1, 2, ..., L_\mu - 1\}$, by one of the following two operations:

1. Swapping either two columns or two rows of the matrix $\mathbf{K}^\mu_j$.
2. Splitting one of the rows (or columns) of $\mathbf{K}^\mu_j$ denoted by v, into two row vectors v_1 and v_2, i.e., $v = [v_1 \;\; v_2]$ (or $v = [v_1 \;\; v_2]'$). Then, replacing that row (or column) with $[v_1 \;\; 0]$ (or $[v_1 \;\; 0]'$), where 0 represents a zero row vector, and inserting another row (or column) equal to $v = [0 \;\; v_2]$ (or $v = [0 \;\; v_2]'$) into the matrix.

It is desired now to prove for any $j \in \{1, ..., L_\mu - 1\}$, that there exist matrices Φ^μ_j and $\bar{\Phi}^\mu_j$ such that $\mathbf{K}^\mu_j = \Phi^\mu_j \mathbf{K}^\mu_{j+1} \bar{\Phi}^\mu_j$.

1. Assume that $\mathbf{K}^\mu_{j+1}$ is derived from $\mathbf{K}^\mu_j$ by swapping its gth and q^{th} columns. It is straightforward to show in this case, that the matrices Φ^μ_j and $\bar{\Phi}^\mu_j$ will be as follows:

(a) Φ_j^μ is an identity matrix, whose dimension is equal to the number of rows of $\mathbf{K}_j^\mu$.

(b) $\bar{\Phi}_j^\mu$ is derived from an identity matrix, whose dimension is equal to the number of columns of $\mathbf{K}_j^\mu$, by setting the (g,g) and (q,q) entries of this identity matrix to zero, and setting its (g,q) and (q,g) entries to one.

It is to be noted that if two rows of $\mathbf{K}_j^\mu$ instead of two columns are swapped, then the procedures to obtain the matrices Φ_j^μ and $\bar{\Phi}_j^\mu$ should also be swapped.

2. Assume that one of the columns of the matrix $\mathbf{K}_j^\mu$ is split into two columns as described before (note that the case of row split can be carried out in a similar manner). For instance, suppose that $\mathbf{K}_j^\mu$ is as follows:

$$\mathbf{K}_j^\mu = \begin{bmatrix} M_1 & m_2 & M_3 \\ M_4 & m_5 & M_6 \end{bmatrix} \tag{8.7}$$

where:

$$\begin{aligned} &M_1 \in \Re^{g_1 \times q_1}, \ m_2 \in \Re^{g_1 \times 1}, \ M_3 \in \Re^{g_1 \times q_2}, \\ &M_4 \in \Re^{g_2 \times q_1}, \ m_5 \in \Re^{g_2 \times 1}, \ M_6 \in \Re^{g_2 \times q_2}, \end{aligned} \tag{8.8}$$

In addition, consider:

$$\mathbf{K}_{j+1}^\mu = \begin{bmatrix} M_1 & m_2 & 0_{g_1 \times 1} & M_3 \\ M_4 & 0_{g_2 \times 1} & m_5 & M_6 \end{bmatrix} \tag{8.9}$$

It can be easily verified that $\Phi_j^\mu = I_{g_1+g_2}$, and:

$$\bar{\Phi}_j^\mu = \begin{bmatrix} I_{q_1} & 0_{q_1 \times 1} & 0_{q_1 \times q_2} \\ 0_{1 \times q_1} & 1 & 0_{1 \times q_2} \\ 0_{1 \times q_1} & 1 & 0_{1 \times q_2} \\ 0_{q_2 \times q_1} & 0_{q_2 \times 1} & I_{q_2} \end{bmatrix} \tag{8.10}$$

Note that $0_{q \times g}$ and I_g represent the $q \times g$ zero matrix and the $g \times g$ identity matrix, respectively, for any $g, q \geq 1$.

Hence, it is shown that for each of the two operations discussed earlier, there exist the matrices Φ_j^μ and $\bar{\Phi}_j^\mu$, which satisfy the aforementioned property. The matrices Φ_μ and $\bar{\Phi}_\mu$ can now be obtained from the following equations:

$$\Phi_\mu = \Phi_1^\mu \Phi_2^\mu \cdots \Phi_{(L_\mu - 1)}^\mu, \quad \bar{\Phi}_\mu = \bar{\Phi}_{(L_\mu - 1)}^\mu \bar{\Phi}_{(L_\mu - 2)}^\mu \cdots \bar{\Phi}_1^\mu \tag{8.11}$$

■

Theorem 1 states that there exist matrices Φ_μ and $\bar{\Phi}_\mu$ for the matrix $\mathbf{K}_\mu$ ($\mu \in \bar{l}$) derived from $\mathbf{K}$ using Procedures 1, 2 and 3, such that they satisfy the equation (8.6). However, since the proof of Theorem 1 relies on a sequence of matrices, the proposed procedure may not be efficient to compute Φ_μ and $\bar{\Phi}_\mu$ for an information flow matrix with a large number of block entries. The following theorem presents a more efficient approach to obtain Φ_μ and $\bar{\Phi}_\mu$.

Theorem 2 *Choose at least one nonzero block entry from each block column and each block row of* $\mathbf{K}_\mu$, $\mu \in \bar{l}$, *and let them be denoted by* $k_{i_1 j_1}, k_{i_2 j_2}, \ldots, k_{i_p j_p}$. *Suppose that* $k_{i_q j_q}$, $q = 1, 2, \ldots, p$, *is the* (i'_q, j'_q) *block entry of the matrix* $\mathbf{K}_\mu$. *Denote the* h_1^{th} *block column of* Φ_μ *and the* h_2^{th} *block row of* $\bar{\Phi}_\mu$ *with* Π_{h_1} *and* $\bar{\Pi}_{h_2}$, *respectively, for* $h_1 = 1, 2, \ldots, m^\mu$, $h_2 = 1, 2, \ldots, r^\mu$ *(note that* m^μ *and* r^μ *are defined in procedure 3). Then:*

$$\Pi_{i'_q} = \begin{bmatrix} 0_{m_1 \times m_{i_q}} \\ 0_{m_2 \times m_{i_q}} \\ \vdots \\ 0_{m_{(i_q-1)} \times m_{i_q}} \\ I_{m_{i_q}} \\ 0_{m_{(i_q+1)} \times m_{i_q}} \\ \vdots \\ 0_{m_\nu \times m_{i_q}} \end{bmatrix}, \quad \bar{\Pi}_{j'_q} = \begin{bmatrix} 0_{r_{j_q} \times r_1} \\ 0_{r_{j_q} \times r_2} \\ \vdots \\ 0_{r_{j_q} \times r_{(j_q-1)}} \\ I_{r_{j_q}} \\ 0_{r_{j_q} \times r_{(j_q+1)}} \\ \vdots \\ 0_{r_{j_q} \times r_\nu} \end{bmatrix}^T \tag{8.12}$$

for any $q \in \{1, 2, \ldots, p\}$.

Proof It is shown in Theorem 1 that the matrices Φ_μ and $\bar{\Phi}_\mu$ exist to satisfy the equation (8.6). As a result, this equation holds for any arbitrary values for the block entries $k_{\sigma_1 \sigma_2}$, $\sigma_1, \sigma_2 \in \bar{\nu}$. Replace all block entries $k_{\sigma_1 \sigma_2}$'s in the equation (8.6), except $k_{i_q j_q}$, with zero matrices. It can be concluded from (8.6) that:

$$\tilde{\mathbf{K}}_{i_q j_q} = \Pi_{i'_q} k_{i_q j_q} \bar{\Pi}_{j'_q} \tag{8.13}$$

where $\tilde{\mathbf{K}}_{i_q j_q}$ is obtained from $\mathbf{K}$ by replacing all of its block entries with zero matrices, except for its (i_q, j_q) block entry $k_{i_q j_q}$. The proof follows immediately from the equation (8.13). ■

It is to be noted that the matrices Φ_μ and $\bar{\Phi}_\mu$ are uniquely determined. To illustrate the method proposed in Theorem 2, consider again the matrix $\mathbf{K}_2$ given by (8.5), which is obtained from $\mathbf{K}$ in (8.4), and assume that the subsystems of the original system are all single-input single-output (SISO). As the first step in computing Φ_2 and $\bar{\Phi}_2$, choose some of the nonzero entries of $\mathbf{K}_2$, such that at least one entry from each column and each row of $\mathbf{K}$ is included. Let these entries be $k_{11}, k_{21}, k_{31}, k_{22}, k_{44}$, and k_{33}. The position of these entries in the matrix $\mathbf{K}_2$ are $(1,1),(2,1),(3,1),(4,2),(5,3),(6,4)$, respectively. Using Theorem 2, one can obtain the matrices Φ_2 and $\bar{\Phi}_2$ for this example, as follows:

$$\Phi_2 = \begin{bmatrix} 1 & 0 & 0 & 0 & 0 & 0 \\ 0 & 1 & 0 & 1 & 0 & 0 \\ 0 & 0 & 1 & 0 & 0 & 1 \\ 0 & 0 & 0 & 0 & 1 & 0 \end{bmatrix}, \quad \bar{\Phi}_2 = \begin{bmatrix} 1 & 0 & 0 & 0 \\ 0 & 1 & 0 & 0 \\ 0 & 0 & 0 & 1 \\ 0 & 0 & 1 & 0 \end{bmatrix} \tag{8.14}$$

It is very easy to verify that these matrices satisfy the relation (8.6).

8.5 Linear time-invariant control law

In this section, it is desired to find conditions for the existence of a stabilizing LTI controller for the system $\mathscr{S}(\mathscr{K})$. Furthermore, a procedure is given to achieve pole placement using a LTI control law. Design of a structurally constrained linear-quadratic optimal controller is then studied.

8.5.1 Pole placement

Definition 2 *Define $\mathscr{S}_\mu$, $\mu \in \bar{1}$, as an interconnected system with the following state-space representation:*

$$\begin{aligned} \dot{\mathbf{x}}_\mu(t) &= A\mathbf{x}_\mu(t) + \mathbf{B}^\mu \mathbf{u}_\mu(t) \\ \mathbf{y}_\mu(t) &= \mathbf{C}^\mu \mathbf{x}_\mu(t) + \mathbf{D}^\mu \mathbf{u}_\mu(t) \end{aligned} \tag{8.15}$$

where the system parameters are related to the state-space matrices of the system $\mathscr{S}$ given by (8.1), as shown below:

$$\mathbf{B}^\mu = B\Phi_\mu, \quad \mathbf{C}^\mu = \bar{\Phi}_\mu C, \quad \mathbf{D}^\mu = \bar{\Phi}_\mu D \Phi_\mu \tag{8.16}$$

$\mathbf{u}_\mu(t) \in \Re^{m^\mu}$ and $\mathbf{y}_\mu(t) \in \Re^{r^\mu}$ are the input and the output of $\mathscr{S}_\mu$, respectively, and $\mathbf{x}_\mu(0) = x(0)$. For any $\mu \in \bar{1}$, define the information flow matrix $\mathscr{K}_\mu$ for the system $\mathscr{S}_\mu$ as a matrix obtained from

$\mathbf{K}_\mu$ *by replacing its nonzero block entry* k_{ij}*, with a* $m_i \times r_j$ *matrix whose entries are all equal to one, for any* $i, j \in \bar{\nu}$.

Theorem 3 *For any* $\mu \in \bar{l}$*, the systems* $\mathscr{S}_\mu(\mathscr{K}_\mu)$ *and* $\mathscr{S}(\mathscr{K})$ *are* analogous *with respect to the set of all LTI controllers.*

Proof Denote the transfer function matrix of any nonzero control component k_{ij} with $K_{ij}(s)$, $i, j \in \bar{\nu}$ (the dimension of $K_{ij}(s)$ is the same as k_{ij} but the function itself is yet to be designed). Replace the block k_{ij} with $K_{ij}(s)$ in the matrices $\mathbf{K}$ and $\mathbf{K}_\mu$ for any $i, j \in \bar{\nu}$, and denote the resultant control transfer function matrices with $K(s)$ and $K_\mu(s)$, respectively. It can be easily concluded from Theorem 1 that:

$$K(s) = \Phi_\mu K_\mu(s)\bar{\Phi}_\mu \tag{8.17}$$

Assume the control transfer function matrix $K(s)$ is such that the matrix $I_r - DK(s)$ is nonsingular. It is known that the state of the system $\mathscr{S}$ under the controller $K(s)$ satisfies the following equation:

$$X(s) = \left(sI_n - A - BK(s)\left(I_r - DK(s)\right)^{-1}C\right)^{-1} x(0) \tag{8.18}$$

On the other hand, it can be easily verified that $I_{r^\mu} - \bar{\Phi}_\mu D\Phi_\mu K_\mu(s)$ is nonsingular due to the assumption $\det(I_r - DK(s)) \neq 0$. Similarly, the state of the system $\mathscr{S}_\mu$ under the controller $K_\mu(s)$ can be obtained as follows:

$$\mathbf{X}_\mu(s) = \left(sI_n - A - \mathbf{B}^\mu K_\mu(s)\left(I_{r^\mu} - \mathbf{D}^\mu K_\mu(s)\right)^{-1}\mathbf{C}^\mu\right)^{-1}\mathbf{x}_\mu(0) \tag{8.19}$$

Furthermore, using the equations (8.16) and (8.17), one can write:

$$\begin{aligned} BK(s)(I_r - DK(s))^{-1}C &= B\Phi_\mu K_\mu(s)\bar{\Phi}_\mu\left(I_r - D\Phi_\mu K_\mu(s)\bar{\Phi}_\mu\right)^{-1}C \\ &= B\Phi_\mu K_\mu(s)\left(I_{r^\mu} - \bar{\Phi}_\mu D\Phi_\mu K_\mu(s)\right)^{-1}\bar{\Phi}_\mu C \\ &= \mathbf{B}^\mu K_\mu(s)\left(I_{r^\mu} - \mathbf{D}^\mu K_\mu(s)\right)^{-1}\mathbf{C}^\mu \end{aligned} \tag{8.20}$$

The proof follows from the relations (8.18), (8.19), and (8.20). ■

Corollary 1 *For any* $\mu \in \bar{l}$*, the systems* $\mathscr{S}_\mu(\mathscr{K}_\mu)$ *and* $\mathscr{S}(\mathscr{K})$ *are* analogous *with respect to the set of all continuous-time* static *controllers.*

Proof The proof is omitted due to its similarity to the proof of Theorem 3. ■

Remark 2 *It can be easily concluded from Theorem 3 and Corollary 1 that all of the systems* $\mathscr{S}(\mathscr{K})$, $\mathscr{S}_1(\mathscr{K}_1), \mathscr{S}_2(\mathscr{K}_2), ..., \mathscr{S}_l(\mathscr{K}_l)$ *are* analogous *with respect to the set of continuous-time dynamic LTI controllers, as well as the set of continuous-time static controllers. As a result, in order to design a continuous-time dynamic (or static) LTI controller for the system* $\mathscr{S}$ *with respect to the information flow structure* $\mathscr{K}$ *to achieve any design objective (such as pole placement), one can equivalently design a continuous-time LTI controller for the system* $\mathscr{S}_\mu$, $\mu \in \bar{l}$, *with respect to the information flow structure* $\mathscr{K}_\mu$, *to attain the same objective. The mapping between the components of* $\mathbf{K}$ *and* $\mathbf{K}_\mu$ *(derived from the equation (8.6)) can then be used to find the corresponding controller for the system* $\mathscr{S}(\mathscr{K})$. *The important advantage of this indirect design procedure is that the information flow structure* $\mathscr{K}_\mu$ *is block diagonal, and hence the problem is converted to the conventional decentralized control design problem, which can be handled by the existing methods [2; 14].*

The question arises now as which of the systems $\mathscr{S}_1, \mathscr{S}_2, ..., \mathscr{S}_l$ is more appropriate to be employed for the aforementioned control design procedure. It is to be noted that all of these systems are *analogous*, and hence possess similar characteristics in terms of output performance. However, a smart choice of system here is of crucial importance in terms of simplifying the control design problem. This will be discussed in detail later.

Partition now the matrices $\mathbf{B}^\mu, \mathbf{C}^\mu$ and $\mathbf{D}^\mu$, $\mu \in \bar{l}$, as follows:

$$\mathbf{B}^\mu = \begin{bmatrix} \mathbf{B}_1^\mu & \mathbf{B}_2^\mu & \cdots & \mathbf{B}_{v_\mu}^\mu \end{bmatrix}, \quad \mathbf{C}^\mu = \begin{bmatrix} \mathbf{C}_1^\mu \\ \mathbf{C}_2^\mu \\ \cdots \\ \mathbf{C}_{v_\mu}^\mu \end{bmatrix}, \quad \mathbf{D}^\mu = \begin{bmatrix} \mathbf{D}_{1,1}^\mu & \cdots & \mathbf{D}_{1,v_\mu}^\mu \\ \vdots & \ddots & \vdots \\ \mathbf{D}_{v_\mu,1}^\mu & \cdots & \mathbf{D}_{v_\mu,v_\mu}^\mu \end{bmatrix} \tag{8.21}$$

where:

$$\mathbf{B}_i^\mu \in \Re^{m_i^\mu}, \quad \mathbf{C}_i^\mu \in \Re^{r_i^\mu}, \quad \mathbf{D}_{ij}^\mu \in \Re^{r_i^\mu \times m_j^\mu} \tag{8.22}$$

for any $i, j \in \{1, 2, ..., v_\mu\}$. It is to be noted that m_i^μ and r_i^μ are defined in Procedure 3.

Theorem 4 *Consider an arbitrary region* $\mathscr{R}$ *in the complex plane. There exists a LTI controller for the system* $\mathscr{S}(\mathscr{K})$ *to place all modes of the resultant closed-loop system inside the region* $\mathscr{R}$, *except for those modes which are DFMs of the system* $\mathscr{S}_\mu$ *with respect to* $\mathscr{K}_\mu$, $\mu \in \bar{l}$.

Proof As pointed out in Remark 2, the systems $\mathscr{S}(\mathscr{K})$ and $\mathscr{S}_\mu(\mathscr{K}_\mu)$ are equivalent in terms of pole placement capabilities. On the other hand, it results from the definition of DFM [1] that all of the modes of the system $\mathscr{S}_\mu(\mathscr{K}_\mu)$ except for its DFMs can be placed arbitrarily by using a proper LTI controller. This completes the proof. ∎

Definition 3 *Define decentralized overlapping fixed modes (DOFM) of $\mathscr{S}(\mathscr{K})$ as those modes of the system $\mathscr{S}$ which are fixed with respect to any dynamic LTI controller with the information flow structure $\mathscr{K}$.*

Theorem 4 states that the DOFMs of $\mathscr{S}(\mathscr{K})$ and the DFMs of $\mathscr{S}_\mu(\mathscr{K}_\mu)$, $\forall\, \mu \in \bar{l}$ are the same. Hence, the DOFMs of $\mathscr{S}(\mathscr{K})$ can be obtained from any of the systems $\mathscr{S}_1(\mathscr{K}_1),...,\mathscr{S}_l(\mathscr{K}_l)$. The following procedure is used to determines the DOFMs of the system $\mathscr{S}(\mathscr{K})$ from the DFMs of the system $\mathscr{S}_\mu(\mathscr{K}_\mu)$, $\mu \in \bar{l}$.

Procedure 4 *Consider any arbitrary g belonging to $\bar{l}$. Let $sp(A)$ denote the set of eigenvalues of A. $\lambda \in sp(A)$ is a DOFM of the system $\mathscr{S}$ with respect to the information flow matrix $\mathscr{K}$, if there exists a permutation of $\{1,2,...,\nu_\mu\}$ denoted by the distinct integers $i_1,i_2,...,i_{\nu_\mu}$, such that the rank of the matrix:*

$$\begin{bmatrix} A-\lambda I_n & \mathbf{B}^\mu_{i_1} & \mathbf{B}^\mu_{i_2} & \cdots & \mathbf{B}^\mu_{i_q} \\ \mathbf{C}^\mu_{i_{q+1}} & \mathbf{D}^\mu_{i_{q+1},i_1} & \mathbf{D}^\mu_{i_{q+1},i_2} & \cdots & \mathbf{D}^\mu_{i_{q+1},i_q} \\ \mathbf{C}^\mu_{i_{q+2}} & \mathbf{D}^\mu_{i_{q+2},i_1} & \mathbf{D}^\mu_{i_{q+2},i_2} & \cdots & \mathbf{D}^\mu_{i_{q+2},i_q} \\ \vdots & \vdots & \vdots & \ddots & \vdots \\ \mathbf{C}^\mu_{i_{\nu_\mu}} & \mathbf{D}^\mu_{i_{\nu_\mu},i_1} & \mathbf{D}^\mu_{i_{\nu_\mu},i_2} & \cdots & \mathbf{D}^\mu_{i_{\nu_\mu},i_q} \end{bmatrix} \tag{8.23}$$

is less than n for some $\mu \in \{0,1,...,\nu_\mu\}$.

Remark 3 *According to Procedure 4, the rank of a set of matrices given in (8.23) should be checked to find out if any of the eigenvalues of the matrix A is a DOFM of the system $\mathscr{S}(\mathscr{K})$. It can be easily verified that the number of these matrices grows exponentially by ν_μ (the number of complete bipartite subgraphs of $\mathscr{G}_\mu$). Therefore, in order to reduce the required computations, it is rather desirable to choose the graph $\mathscr{G}_\mu$ from the set of graphs $\{\mathscr{G}_1,...,\mathscr{G}_l\}$, such that it has the minimum number of complete bipartite subgraphs. Moreover, if there is more than one such candidate, the one with fewer number of vertices is more preferable.*

Corollary 2 *The system $\mathscr{S}(\mathscr{K})$ is stabilizable by means of a dynamic LTI controller if and only if it does not have any DOFM in the closed right-half plane with respect to the information flow matrix $\mathscr{K}$.*

Proof The proof follows immediately from Theorem 4. ■

The following theorem presents a method to characterize the DOFMs of the system $\mathscr{S}(\mathscr{K})$ in terms of the transmission zeros of a set of systems.

Theorem 5 *$\lambda \in sp(A)$ is a DOFM of the system $\mathscr{S}(\mathscr{K})$ if and only if it is a transmission zero of the following system:*

$$\begin{aligned}\dot{\mathbf{x}}(t) &= A\mathbf{x}(t) + \begin{bmatrix} \mathbf{B}^l_{i_1} & \mathbf{B}^l_{i_2} & \cdots & \mathbf{B}^l_{i_q} \end{bmatrix} \mathbf{u}(t) \\ \mathbf{y}(t) &= \begin{bmatrix} \mathbf{C}^l_{i_1} \\ \mathbf{C}^l_{i_2} \\ \vdots \\ \mathbf{C}^l_{i_q} \end{bmatrix} \mathbf{x}(t) + \begin{bmatrix} 0 & \mathbf{D}^l_{i_1 i_2} & \cdots & \mathbf{D}^l_{i_1 i_q} \\ \mathbf{D}^l_{i_2 i_1} & 0 & \cdots & \mathbf{D}^l_{i_2 i_q} \\ \vdots & \vdots & \ddots & \vdots \\ \mathbf{D}^l_{i_q i_1} & \mathbf{D}^l_{i_q i_2} & \cdots & 0 \end{bmatrix} \mathbf{u}(t)\end{aligned} \tag{8.24}$$

for any $q \in \{1, ..., \nu_l\}$ and any arbitrary subset $\{i_1, i_2, ..., i_q\}$ of $\{1, 2, ..., \nu_l\}$.

Proof It was shown earlier that the DOFMs of the system $\mathscr{S}(\mathscr{K})$ are the same as the DFMs of the system $\mathscr{S}_l(\mathscr{K}_l)$. Furthermore, since the matrix $\mathscr{K}_l$ is diagonal (because the graph $\mathscr{G}_l$ is composed of some disjoint edges), it results from [2] that the DFMs of the system $\mathscr{S}_l(\mathscr{K}_l)$ are the same as the common transmission zeros of the systems given by (8.24). This completes the proof. ■

Remark 4 *The results of Theorem 5 are obtained in [2] for the particular case when the information flow matrix $\mathscr{K}$ is block diagonal. Furthermore, the system given by (8.24) is constructed by using Kronecker product in [2], while it is formed by means of graph theory in this chapter. Therefore, Theorem 5 presents the results for the most general information flow structure compared to the ones given in [2].*

8.5.2 Optimal LTI controller

Assume that the system $\mathscr{S}$ is stabilizable with respect to the information flow matrix $\mathscr{K}$ by means of a dynamic LTI controller. It is desired to find a LTI controller with the zero initial state and the transfer function matrix $K(s)$ corresponding to the information flow structure $\mathscr{K}$, such that it minimizes the following LQR performance index:

$$J := \int_0^\infty \left(x(t)^T Q x(t) + u(t)^T R u(t)\right) dt \tag{8.25}$$

where $R \in \Re^{m\times m}$ and $Q \in \Re^{n\times n}$ are positive definite and positive semi-definite matrices, respectively, and where:

$$u(t) = \left[\begin{array}{cccc} u_1(t)^T & u_2(t)^T & \cdots & u_v(t)^T \end{array}\right]^T \tag{8.26}$$

Lemma 1 *The matrix Φ_1 corresponding to the information flow matrix $\mathscr{K}_1$ is equal to I_m.*

Proof : The proof follows directly from the procedure of constructing the graph $\mathscr{G}_1$ and Theorem 2. ■

The transfer function matrix $K_1(s)$ constructed in terms of $K(s)$ in the proof of Theorem 3 (for $\mu = 1$) will be used in the next Theorem.

Theorem 6 *Consider the systems $\mathscr{S}$ and $\mathscr{S}_1$ under the controllers $K(s)$ and $K_1(s)$, respectively. Assume that the matrix $I_r - DK(s)$ is nonsingular. Then $J = J_1$, where:*

$$J_1 := \int_0^\infty \left(\mathbf{x}_1(t)^T Q \mathbf{x}_1(t) + \mathbf{u}_1(t)^T R \mathbf{u}_1(t)\right) dt \tag{8.27}$$

Proof It follows from the proof of Theorem 3 (with $\mu = 1$) that $x(t) = \mathbf{x}_1(t)$ for all $t \geq 0$. Besides, it results from this equality and $\Phi_1 = I_m$ (Lemma 1), that $u(t) = \mathbf{u}_1(t)$ for all $t \geq 0$. This completes the proof. ■

Theorem 6 states that in order to find a controller $K(s)$ which minimizes the performance index (8.25) for the system $\mathscr{S}$ while it meets the information flow constraint given by $\mathscr{K}$, one can equivalently pursue the following two steps:

1. Design the decentralized LTI controller $K_1(s)$ in such a way that it minimizes the performance index (8.27) for the system $\mathscr{S}_1(\mathscr{K}_1)$.

2. Find the controller $K(s)$ from the relation $K(s) = \Phi_1 K_1(s) \bar{\Phi}_1$.

It is to be noted that the decentralized optimal control design problem has been studied intensively in the literature, and a number of approaches for obtaining an optimal or a near-optimal decentralized controller are given accordingly, e.g., see [15; 16; 17].

8.6 Non-LTI control law

In this section, the procedure of designing different types of controllers, such as periodic or sampled-data control laws, for the system $\mathscr{S}(\mathscr{K})$ is investigated. Moreover, a necessary and sufficient condition for the stabilizability of the system $\mathscr{S}(\mathscr{K})$ is given. To develop the remaining results of this work, it is hereafter assumed that the system $\mathscr{S}$ is strictly proper, i.e. $D = 0$.

8.6.1 Generalized sampled-data hold function

Periodic control design using generalized sampled-data hold function (GSHF) and its advantages have been studied intensively in the literature [18; 19; 20]. Assume that it is desired to design a GSHF for the system $\mathscr{S}$, which complies with the information flow structure $\mathscr{K}$. Let this GSHF be denoted by $F(t)$. Hence, the hold controller will be as follows:

$$u(t) = F(t)y[\kappa], \quad \kappa h \leq t < (\kappa+1)h, \quad \kappa \geq 0 \tag{8.28}$$

where h represents the sampling periodic. Note that the discrete argument corresponding to the samples of any signal is enclosed in brackets (e.g., $y[\kappa] := y(\kappa h)$).

Theorem 7 *The systems $\mathscr{S}(\mathscr{K}), \mathscr{S}_1(\mathscr{K}_1), \dots, \mathscr{S}_l(\mathscr{K}_l)$ are all* analogous *with respect to the set of all hold controllers (GSHFs).*

Proof To prove the theorem, it suffices to show that $\mathscr{S}(\mathscr{K})$ and $\mathscr{S}_\mu(\mathscr{K}_\mu)$ are *analogous* with respect to all hold controllers, for any $\mu \in \bar{l}$. Consider a GSHF $F(t)$ which complies with the information flow structure $\mathscr{K}$. Utilize the proper transformation on $F(t)$ to obtain the equivalent hold function $F_\mu(t)$ for the system $\mathscr{S}_\mu(\mathscr{K}_\mu)$. Note that $F_\mu(t)$ can be attained using the mapping between the components of $\mathbf{K}$ and $\mathbf{K}_\mu$ (see Remark 1). Since $F(t)$ and $F_\mu(t)$ comply with the information flow matrices $\mathscr{K}$ and $\mathscr{K}_\mu$, respectively, it is straightforward to show that $F(t) = \Phi_\mu F_\mu(t)\bar{\Phi}_\mu$. On the other hand, it follows from (8.28) that:

$$\dot{x}(t) = Ax(t) + BF(t)Cx[\kappa] \tag{8.29}$$

and consequently:

$$\begin{aligned} \dot{\mathbf{x}}_\mu(t) &= A\mathbf{x}_\mu(t) + \mathbf{B}^\mu F_\mu(t)\mathbf{C}^\mu \mathbf{x}_\mu[\kappa] \\ &= A\mathbf{x}_\mu(t) + B\Phi_\mu F_\mu(t)\bar{\Phi}_\mu C\mathbf{x}_\mu[\kappa] \\ &= A\mathbf{x}_\mu(t) + BF(t)C\mathbf{x}_\mu[\kappa] \end{aligned} \tag{8.30}$$

for all $t \in [\kappa h, (\kappa+1)h)$, $\kappa \geq 0$. The equations (8.29) and (8.30), and the equality $x(0) = \mathbf{x}_\mu(0)$ result in the relation $x(t) = \mathbf{x}_\mu(t)$ for all $t \geq 0$. Conversely, for any GSHF $F_\mu(t)$ complying with the information flow matrix $\mathscr{K}_\mu$, it is straightforward to show that the state of the system $\mathscr{S}$ under the GSHF $F(t) = \Phi_\mu F_\mu(t)\bar{\Phi}_\mu$ is identical to that of the system $\mathscr{S}_\mu$ under $F_\mu(t)$. ∎

Theorem 7 states that the problem of designing a GSHF for the system $\mathscr{S}(\mathscr{K})$ can be formulated as the problem of designing a GSHF for the system $\mathscr{S}_\mu(\mathscr{K}_\mu)$ for any $\mu \in \bar{l}$. However, due to the decentralized structure of the control for $\mathscr{S}_\mu(\mathscr{K}_\mu)$, $\mu \in \bar{l}$, the corresponding GSHF design can be accomplished by using the existing methods [21; 22].

8.6.2 Sampled-data controller

A typical sampled-data controller consists of a sampler, a zero-order hold (ZOH) and a discrete-time controller. It is to be noted that a sampled-data controller acts as a time-varying control law for the continuous-time system. It is desired in this subsection to present a method for designing a sampled-data controller for the system $\mathscr{S}$, whose structure complies with a given information flow matrix $\mathscr{K}$. Throughout the remainder of this chapter, the term *linear shift-invariant* (LSI) will be used instead of LTI, for discrete-time systems.

Theorem 8 *The systems $\mathscr{S}(\mathscr{K}), \mathscr{S}_1(\mathscr{K}_1), \ldots, \mathscr{S}_l(\mathscr{K}_l)$ are all* analogous *with respect to the set of all LSI sampled-data controllers.*

Proof Denote the sampling period with h, and the discrete-time equivalent models of the systems $\mathscr{S}, \mathscr{S}_1, \ldots, \mathscr{S}_l$ with $\bar{\mathscr{S}}, \bar{\mathscr{S}}_1, \ldots, \bar{\mathscr{S}}_l$, respectively. Assume that the system $\bar{\mathscr{S}}$ is represented by:

$$\begin{aligned} x[\kappa+1] &= \bar{A}x[\kappa] + \bar{B}u[\kappa] \\ y[\kappa] &= Cx[\kappa] \end{aligned} \tag{8.31}$$

Similarly, let the system $\bar{\mathscr{S}}_\mu$ be represented by:

$$\begin{aligned} \mathbf{x}_\mu[\kappa+1] &= \bar{A}\mathbf{x}_\mu[\kappa] + \bar{\mathbf{B}}^\mu \mathbf{u}_\mu[\kappa] \\ \mathbf{y}_\mu[\kappa] &= \mathbf{C}^\mu \mathbf{x}_\mu[\kappa], \qquad \mu \in \bar{l} \end{aligned} \tag{8.32}$$

It can be easily verified that:

$$\bar{\mathbf{B}}^\mu = \int_0^h e^{\tau A}\mathbf{B}^\mu \, d\tau = \int_0^h e^{\tau A} B \, d\tau \times \Phi_\mu = \bar{B}\Phi_\mu \tag{8.33}$$

It results from (8.31), (8.32), and (8.33) that the state-space matrices of $\bar{\mathscr{S}}$ are related to those of $\bar{\mathscr{S}}_\mu$, exactly the same way the state-space matrices of $\mathscr{S}$ and $\mathscr{S}_\mu$ are related. Hence, the systems $\bar{\mathscr{S}}$ and $\bar{\mathscr{S}}_\mu$ are *analogous* with respect to the LSI controllers. Consider now a discrete-time LSI controller with the transfer function matrix $\bar{K}(z)$ for the system $\bar{\mathscr{S}}(\mathscr{K})$. Construct a discrete-time LSI controller with the transfer function matrix $\bar{K}_\mu(z)$ for the system $\bar{\mathscr{S}}_\mu(\mathscr{K}_\mu)$, such that it corresponds to the controller $\bar{K}(z)$ for $\bar{\mathscr{S}}(\mathscr{K})$. This controller can be obtained from the mapping between the components of $\mathbf{K}$ and $\mathbf{K}_\mu$. It is straightforward to show that $\bar{K}(z) = \Phi_\mu \bar{K}_\mu(z)\bar{\Phi}_\mu$. Applying the controller $\bar{K}(z)$ to the system $\bar{\mathscr{S}}$ and the controller $\bar{K}_\mu(z)$ to $\bar{\mathscr{S}}_\mu$, one can conclude (using an approach similar to the one given in the proof of Theorem 3) that $x[\kappa] = \mathbf{x}_\mu[\kappa]$ and $u[\kappa] = \Phi_\mu \mathbf{u}_\mu[\kappa]$ for any $\kappa \geq 0$. Therefore,

$$\begin{aligned}
x(t) &= e^{(t-\kappa h)A}x[\kappa] + \int_{\kappa h}^{t} e^{(\tau-\kappa h)A}Bu[\kappa]\,d\tau \\
&= e^{(t-\kappa h)A}\mathbf{x}_\mu[\kappa] + \int_{\kappa h}^{t} e^{(\tau-\kappa h)A}B\Phi_\mu \mathbf{u}_\mu[\kappa]\,d\tau \\
&= e^{(t-\kappa h)A}\mathbf{x}_\mu[\kappa] + \int_{\kappa h}^{t} e^{(\tau-\kappa h)A}\mathbf{B}^\mu \mathbf{u}_\mu[\kappa]\,d\tau \\
&= \mathbf{x}_\mu(t)
\end{aligned} \tag{8.34}$$

for any $t \in [\kappa h, (\kappa+1)h)$, $k \geq 0$. Similarly, it can be easily verified that given any controller $\bar{K}_\mu(z)$ for the system $\bar{\mathscr{S}}_\mu(\mathscr{K})$, the controller $\bar{K}(z) := \Phi_\mu \bar{K}_\mu(z)\bar{\Phi}_\mu$ corresponds to the information flow matrix $\mathscr{K}$. Moreover, the state of the system $\mathscr{S}$ under the controller $\bar{K}(z)$ is the same as that of $\mathscr{S}_\mu$ under $\bar{K}_\mu(z)$. ■

It is assumed in the proof of Theorem 8 that $D = 0$. However, its results can be easily extended to the case when $D \neq 0$. Note that finding a sampled-data decentralized control law to achieve certain design objectives has been investigated in the literature, e.g, see [23].

8.6.3 Finite-dimensional linear time-varying controller

It is well-known that finite-dimensional linear time-varying (LTV) controllers are superior to their LTI counterparts in many control applications [21]. It is desired in this subsection to present a procedure for designing a finite-dimensional LTV controller complying with the information flow matrix $\mathscr{K}$, for the system $\mathscr{S}$. Note that throughout this work, the term "finite-dimensional LTV controller" refers to a control law which can be represented by the following state-space model:

$$\dot{\tilde{x}}(t) = \tilde{A}(t)\tilde{x}(t) + \tilde{B}(t)\tilde{u}(t)$$
$$\tilde{y}(t) = \tilde{C}(t)\tilde{x}(t) + \tilde{D}(t)\tilde{u}(t) \tag{8.35}$$

Theorem 9 *The systems $\mathscr{S}(\mathscr{K}), \mathscr{S}_1(\mathscr{K}_1), ..., \mathscr{S}_l(\mathscr{K}_l)$ are all* analogous *with respect to the set of all finite-dimensional LTV controllers.*

The foregoing theorem extends the results of Theorem 3 to the case when the controllers are finite-dimensional LTV (as opposed to LTI). The proof of Theorem 9 is similar to that of Theorem 3 (but should be carried out in the time-domain). The details of the proof are omitted here. However, the statement that $\mathscr{S}(\mathscr{K})$ and $\mathscr{S}_\mu(\mathscr{K}_\mu)$, $\mu \in \bar{l}$, are *analogous* with respect to all finite-dimensional LTV controllers can be intuitively justified as follows:

One can easily verify by using the comments given in the proofs of Theorems 1 and 2, that $\mathbf{B}^\mu$ *is derived from B by rearranging its columns and repeating some of them (repetition results from the fact that some of the vertices in set 1 of the graph $\mathscr{G}$ have recurred to construct the graph $\mathscr{G}_\mu$). Analogously,* $\mathbf{C}^\mu$ *is derived from C by rearranging its rows and repeating some of them. These repetitions and rearrangements and their interpretations are described below:*

1. *Repetition of the rows of C indicates that some of the outputs of $\mathscr{S}$ are duplicated to construct the system $\mathscr{S}_\mu$. To justify the necessity of this recurrence, assume that one output of the system $\mathscr{S}$ contributes to two different control inputs. This means that the corresponding control agent is not localized, and hence the corresponding information flow structure is not decentralized. However, by duplicating this output of the system to create a redundant output, and by applying the two resulting outputs to the two above-mentioned control inputs, the resultant control structure will be decentralized, while its functionality is essentially equivalent to the original control system.*

2. *Regarding the repetition in the columns of* $\mathbf{B}^\mu$, *assume that two outputs of the system contribute to one control agent. Since the controller is linear, one can split the control agent to two sub-agents such that each of the two outputs of the system goes to one of these sub-agents. The control signal of the original control agent is, in fact, equal to the summation of the control signals of these two sub-agents (this results from the principle of superposition). Again, the*

functionality of the resultant control system is equivalent to the original one, while its structure is decentralized.

3. *The rearrangement of the rows and the columns of C and B is equivalent to the reordering of the inputs and the outputs of $\mathscr{S}$, and has no impact on the operation of the overall control system.*

Taking the aforementioned interpretations into consideration, the system $\mathscr{S}_\mu$ is indeed constructed from $\mathscr{S}$ in such a way that the control structure $\mathscr{K}$ is converted to a decentralized structure $\mathscr{K}_\mu$, while essentially both control systems perform identically.

Theorem 9 implies that to design a finite-dimensional LTV controller for the system $\mathscr{S}(\mathscr{K})$, one can first design a LTV controller for one of the systems $\mathscr{S}_1(\mathscr{K}_1),...,\mathscr{S}_l(\mathscr{K}_l)$. This result will be exploited in the following section to present one of the main contributions of the present work.

8.6.4 General controller

The objective of this subsection is to find out under what conditions the system $\mathscr{S}(\mathscr{K})$ is stabilizable by means of a general control law (i.e. nonlinear and time-varying), when there exists no stabilizing LTI controller.

Theorem 10 *The systems $\mathscr{S}(\mathscr{K})$ and $\mathscr{S}_1(\mathscr{K}_1)$ are* analogous *with respect to any type of controller (i.e. nonlinear or time-varying).*

Proof As pointed out in the discussion following Theorem 9, the configurations of the systems $\mathscr{S}$ and $\mathscr{S}_1$ are essentially equivalent. In other words, the system $\mathscr{S}_1$ is obtained from $\mathscr{S}$ by introducing some redundant outputs or control agents and reordering them, in such a way that the information flow structure $\mathscr{K}$ is converted to $\mathscr{K}_1$. Note that according to Lemma 1, $B = \mathbf{B}^1$. Hence, the state of the closed-loop system corresponding to the pair $(\mathscr{S}, \mathscr{K})$ is identical to that of the pair $(\mathscr{S}_1, \mathscr{K}_1)$, regardless of the type of the control law. ■

It is to be noted that unlike $\mathscr{S}(\mathscr{K})$ and $\mathscr{S}_1(\mathscr{K}_1)$, the systems $\mathscr{S}(\mathscr{K})$ and $\mathscr{S}_\mu(\mathscr{K}_\mu)$, $\mu \in \{2,3,...,l\}$, are not *analogous* with respect to any type of controller, in general. This results from the fact that the superposition principle presented in item 2 of the discussion following Theorem 9 does not apply here, as the controllers are nonlinear.

Remark 5 *It follows immediately from Theorem 10 that the system $\mathscr{S}(\mathscr{K})$ is stabilizable if and only if the system $\mathscr{S}_1(\mathscr{K}_1)$ is stabilizable.*

It is shown in [3] that a system is stabilizable with respect to a block-diagonal information flow matrix (i.e. decentralized control structure) if and only if the system does not any unstable quotient fixed mode (QFM). However, QFM is only defined for decentralized control structures. In the following, this notion is extended to the general information flow structure and its property is investigated accordingly.

Definition 4 *$\lambda \in sp(A)$ is a quotient overlapping fixed mode (QOFM) of the system $\mathscr{S}$ with respect to the information flow matrix $\mathscr{K}$, if λ cannot be eliminated by using any type of controller complying with the structure of $\mathscr{K}$.*

Theorem 11 *The QOFMs of the system $\mathscr{S}(\mathscr{K})$ are the same as the QFMs of the system $\mathscr{S}_\mu(\mathscr{K}_\mu)$, $\forall \mu \in \bar{l}$.*

Proof It follows from Theorem 10 that the QOFMs of the system $\mathscr{S}(\mathscr{K})$ are the same as the QFMs of the system $\mathscr{S}_1(\mathscr{K}_1)$. To complete the proof, it suffices to show that the QFMs of the system $\mathscr{S}_1(\mathscr{K}_1)$ are the same as those of the system $\mathscr{S}_\mu(\mathscr{K}_\mu)$, for $\mu = 2, 3, ..., l$. This can be deduced from the following argument:

- The systems $\mathscr{S}_1(\mathscr{K}_1), ..., \mathscr{S}_l(\mathscr{K}_l)$ all have the same A-matrix, and hence the same modes.
- It is shown in [3; 24] that all of the non-QFMs of any system can be eliminated by using a proper finite-dimensional LTV controller.
- Theorem 9 states that the systems $\mathscr{S}_1(\mathscr{K}_1), ..., \mathscr{S}_l(\mathscr{K}_l)$ are all *analogous* to each other with respect to finite-dimensional LTV controllers.

■

Corollary 3 *The system $\mathscr{S}(\mathscr{K})$ is stabilizable if and only if it does not have any unstable QOFM.*

Proof The proof follows immediately from Remark 5 and Theorem 11. ■

8.7 Comparison with existing methods

8.7.1 Comparison with the work presented in [8]

Consider the system $\mathscr{S}$ with the following state-space representation:

$$\dot{x}(t) = Ax + B_1 u_1(t) + B_2 u_2(t) \tag{8.36}$$

where $A \in \Re^{3\zeta \times 3\zeta}$, and

$$B_1 = \begin{bmatrix} B_{11} \\ 0_{\zeta\times\zeta_1} \\ 0_{\zeta\times\zeta_1} \end{bmatrix}, \; B_2 = \begin{bmatrix} 0_{\zeta\times\zeta_2} \\ 0_{\zeta\times\zeta_2} \\ B_{32} \end{bmatrix}, \; x(t) = \begin{bmatrix} x_1(t) \\ x_2(t) \\ x_3(t) \end{bmatrix} \tag{8.37}$$

and $x_i(t) \in \Re^{\zeta}$, $i = 1,2,3$. The outputs of this system are assumed to be the same as its state variables. It is desired now to design a stabilizing structurally constrained static controller for $\mathscr{S}$ with the information flow matrix :

$$\mathscr{K} = \begin{bmatrix} 1 & 1 & 0 \\ 0 & 1 & 1 \end{bmatrix} \tag{8.38}$$

This decentralized overlapping problem is investigated in [8], where the expansion approach is used to solve the problem (see [4] for a numerical version of this example). In this method, the system $\mathscr{S}$ is converted to another system, which is referred to as the expanded system. Subsequently, it is stated that if the expanded system can be stabilized, then the system $\mathscr{S}$ is stabilizable as well. However, since the expanded system is inherently uncontrollable, this approach might be inefficient. It is desired now to demonstrate the effectiveness of the method proposed in this chapter for this example. Using the proposed method, one can easily conclude that the DOFMs of the system $\mathscr{S}$ consist of unobservable modes, uncontrollable modes, and any mode λ for which at least one of the following two matrices:

$$\begin{bmatrix} A-\lambda I_n & B_1 \\ H_1 & o_{2\zeta\times\zeta_1} \end{bmatrix}, \; \begin{bmatrix} A-\lambda I_n & B_2 \\ H_2 & o_{2\zeta\times\zeta_2} \end{bmatrix} \tag{8.39}$$

loses rank, where $H_1 = \begin{bmatrix} I_{2\zeta} & 0_{2\zeta\times\zeta} \end{bmatrix}$ and $H_2 = \begin{bmatrix} 0_{2\zeta\times\zeta} & I_{2\zeta} \end{bmatrix}$. Assume now that the system does not have any unstable DOFM. One can use Procedures 1, 2 and 3 to obtain the matrix $\mathbf{K}_2$ as follows:

$$\mathbf{K}_2 = \begin{bmatrix} k_{11} & 0_{\zeta_1\times\zeta} & 0_{\zeta_1\times\zeta} \\ 0_{\zeta_1\times\zeta} & k_{12} & 0_{\zeta_1\times\zeta} \\ 0_{\zeta_2\times\zeta} & k_{22} & 0_{\zeta_2\times\zeta} \\ 0_{\zeta_2\times\zeta} & 0_{\zeta_2\times\zeta} & k_{23} \end{bmatrix} \tag{8.40}$$

which corresponds to the system $\mathscr{S}_2$ with the following state-space representation:

$$\dot{\mathbf{x}}_2(t) = A\mathbf{x}_2(t) + \mathbf{B}_1^2\mathbf{u}_1^2(t) + \mathbf{B}_2^2\mathbf{u}_2^2(t) + \mathbf{B}_3^2\mathbf{u}_3^2(t) \tag{8.41}$$

where:

$$\mathbf{x}_2(t) = \begin{bmatrix} \mathbf{x}_1^2(t) \\ \mathbf{x}_2^2(t) \\ \mathbf{x}_3^2(t) \end{bmatrix}, \quad \mathbf{B}_1^2 = B_1, \quad \mathbf{B}_2^2 = \begin{bmatrix} B_1 & B_2 \end{bmatrix}, \quad \mathbf{B}_3^2 = B_2 \tag{8.42}$$

and $\mathbf{x}_i^2(t) \in \Re^{\zeta}$, $i = 1,2,3$. It can be concluded from Corollary 1 that designing a *static* structurally constrained controller for the system $\mathscr{S}$ is identical to designing a *static* decentralized controller for the system $\mathscr{S}_2$ (i.e., $\mathbf{u}_i^2(t)$ is constructed in terms of $\mathbf{x}_i^2(t)$ for $i = 1,2,3$). The latter problem can be solved by using either the LMI method proposed in [8] (where this example is presented), or other existing methods, e.g. [15].

8.7.2 Comparison with the work presented in [25]

A method is proposed in [25] for strictly proper systems to determine whether the system is stabilizable with respect to a given information flow matrix by means of LTI controllers. However, this method has the following deficiencies compared to the present work:

1. It cannot be extended to the general proper systems.

2. It translates the stabilizability of a system by means of LTI controllers to that of another system which is, in fact, $\mathscr{S}_1$. However, this may require that the ranks of a huge number of matrices to be checked in order to find out whether the system is stabilizable. For instance, assume that the system $\mathscr{S}$ is composed of 100 SISO subsystems, and that the corresponding information flow matrix $\mathscr{K}$ is a 100×100 matrix, with the first entries of the odd rows and the last entries of

the even rows all equal to zero, and the remaining entries all equal to 1. It is straightforward to show that the number of matrices whose ranks need to be checked by using the method given in [25] is equal to 2^{100}, while the system $\mathscr{S}_2$ can be constructed in such a way that the number of matrices whose ranks need to be checked is equal to 2^2. This sizable difference demonstrates the efficiency of the present work.

8.8 Numerical examples

Example 1 Consider the system $\mathscr{S}$ consisting of three SISO subsystems with the following state-space matrices:

$$A = \begin{bmatrix} 1 & 0 & 1 \\ 0 & 1 & 2 \\ 1 & 2 & 3 \end{bmatrix}, \quad B = \begin{bmatrix} 0 & 1 & 0 \\ 0 & 0 & 1 \\ 1 & 0 & 1 \end{bmatrix}, \quad C = \begin{bmatrix} 1 & 0 & 0 \\ 0 & 0 & 1 \\ 1 & 2 & 2 \end{bmatrix}, \quad D = 0_{3\times 3} \tag{8.43}$$

Consider the information flow matrix $\mathscr{K} = \begin{bmatrix} 1 & 0 & 1 \\ 0 & 1 & 1 \\ 1 & 1 & 1 \end{bmatrix}$, which corresponds to the following control structure:

$$\mathbf{K} = \begin{bmatrix} k_{11} & 0 & k_{13} \\ 0 & k_{22} & k_{23} \\ k_{31} & k_{32} & k_{33} \end{bmatrix} \tag{8.44}$$

This is, in fact, a BAS (or BBD) controller [8; 9]. The following matrix $\mathbf{K}_2$ can be obtained using Procedures 1, 2 and 3:

$$\mathbf{K}_2 = \begin{bmatrix} k_{22} & k_{23} & 0 & 0 \\ k_{32} & k_{33} & 0 & 0 \\ 0 & 0 & k_{11} & 0 \\ 0 & 0 & k_{31} & 0 \\ 0 & 0 & 0 & k_{13} \end{bmatrix} \tag{8.45}$$

Note that for this particular example, $\mathscr{G}_2$ is the best candidate in terms of the subsequent computational complexity. The matrices Φ_2 and $\bar{\Phi}_2$ can be obtained from Theorem 2 as follows:

$$\Phi_1 = \begin{bmatrix} 0 & 0 & 1 & 0 & 1 \\ 1 & 0 & 0 & 0 & 0 \\ 0 & 1 & 0 & 1 & 0 \end{bmatrix}, \quad \Phi_2 = \begin{bmatrix} 0 & 1 & 0 \\ 0 & 0 & 1 \\ 1 & 0 & 0 \\ 0 & 0 & 1 \end{bmatrix} \tag{8.46}$$

Now, the system $\mathscr{S}_2$ can be easily constructed by using the equations (8.15), (8.16) and (8.46). It is desired now to design a structurally constrained controller $K(s)$ for the system $\mathscr{S}$ to achieve a settling time of 4 seconds. and an overshoot of less than 4.5%. It can be easily verified that these design specifications will be met by placing the dominant poles of the closed-loop system at $-1 \pm 1i$. From Procedure 4, it is known that the system $\mathscr{S}$ does not have any DOFMs with respect to the information flow matrix $\mathscr{K}$. Now, using any decentralized pole placement method, e.g., the one proposed in [2] or [14], one can place the dominant poles of the closed-loop system $\mathscr{S}_2$ at $-1 \pm 1i$, as discussed in Remark 2. For instance, using the method given in [2], the following control transfer functions are obtained:

$$\begin{aligned}
&K_{11}(s) = K_{13}(s) = K_{31}(s) = 1 \\
&K_{22}(s) = (-89900 - 96100s - 34100s^2 - 5480s^3 - 409s^4 - 11.5s^5)/\text{Den}(s) \\
&K_{23}(s) = (-15700 - 20500s - 8810s^2 - 1730s^3 - 160s^4 - 5.69s^5)/\text{Den}(s) \\
&K_{32}(s) = (-64500 - 52500s - 16900s^2 - 2740s^3 - 220s^4 - 7.05s^5)/\text{Den}(s) \\
&K_{33}(s) = (-88000 - 64500s - 19200s^2 - 2880s^3 - 219s^4 - 6.7s^5)/\text{Den}(s)
\end{aligned} \tag{8.47}$$

where:

$$\text{Den}(s) = 0.18s^6 + 9.95s^5 + 210.44s^4 + 2269.3s^3 + 13396s^2 + 41488s^1 + 53000 \tag{8.48}$$

(the transfer function of the control component k_{ij} is represented by $K_{ij}(s)$). It is to be noted that using the above control law, the other poles of the closed-loop system will be located at $-4, -6, -7$ and -8.

Example 2 Consider the system $\mathscr{S}$ consisting of four SISO subsystems with the following state-space matrices:

$$A = \begin{bmatrix} 1 & 0 & 0 & 0 & 0 & 0 \\ 0 & -2 & 0 & 0 & 0 & 0 \\ 0 & 0 & -3 & 0 & 0 & 0 \\ 1 & 1 & 1 & 1 & 0 & 0 \\ 1 & 1 & 1 & 0 & -2 & 0 \\ 1 & 1 & 1 & 0 & 0 & -3 \end{bmatrix}, B = \begin{bmatrix} 1 & 0 & 0 & 0 \\ 0 & 3 & 0 & 0 \\ 0 & 1 & 0 & 0 \\ 0 & 0 & 1 & 0 \\ 0 & 0 & 0 & 3 \\ 0 & 0 & 0 & 1 \end{bmatrix}, C = \begin{bmatrix} 0 & 1 & 0 & 0 \\ 0 & 0 & 1 & 0 \\ 0 & 1 & -4 & 0 \\ 1 & 0 & 0 & 0 \\ 0 & 0 & 0 & 1 \\ 1 & 0 & 0 & -4 \end{bmatrix}^T \tag{8.49}$$

and $D = 0_{4\times 4}$. Assume that the information flow matrix for this system is given as follows:

$$\mathscr{K} = \begin{bmatrix} 0 & 0 & 1 & 0 \\ 0 & 1 & 0 & 1 \\ 0 & 0 & 0 & 1 \\ 1 & 0 & 0 & 0 \end{bmatrix} \tag{8.50}$$

One can find the matrices Φ_1 and $\bar{\Phi}_1$ (using Procedures 1, 2 and 3, and Theorem 2), from which one yields that the system $\mathscr{S}$ has two identical DOFMs at $\lambda = +1$ with respect to the information flow matrix $\mathscr{K}$ given by (8.50). Therefore, this system cannot be stabilized by means of a structurally constrained LTI controller. On the other hand, it can be easily verified by using the system $\mathscr{S}_1$ that $\mathscr{S}$ does not have any QOFMs. Hence, this system can be stabilized by means of a constrained LTV controller. Using the method given in [23], one can design a constrained stabilizing sampled-data controller for the system $\mathscr{S}(\mathscr{K})$. Consider a sampling period h equal to 1. The components of the controller will be as follows:

$$\begin{aligned} &\bar{K}_{22}(z) = \bar{K}_{34}(z) = 0, \quad \bar{K}_{13}(z) = \bar{K}_{24}(z) = 1, \\ &\bar{K}_{41}(z) = \left(3945z^5 - 8674z^4 + 1388z^3 + 116.2z^2 - 12.8z - 1.139\right) \\ &\qquad \times \left(z^6 + 2.758z^5 + 877.1z^4 - 1822z^3 + 87.78z^2 + 24.71z + 0.1927\right)^{-1} \end{aligned} \tag{8.51}$$

where $\bar{K}_{ij}(z)$ represents the transfer function of the discrete-time LSI controller corresponding to k_{ij}.

8.9 Conclusions

This work tackles the control design problem for systems with constrained control structure. It is shown that certain modes of the system can be placed freely by means of a linear time-invariant (LTI) structurally constrained controller. The notion of a decentralized overlapping fixed mode (DOFM) is introduced to classify such modes, and an analytical method is given to identify them. In addition, it is shown that the system is stabilizable by means of a LTI structurally constrained controller, if and only if it does not have any unstable DOFM. Furthermore, a graph-theoretic algorithm is proposed to convert the structurally constrained control design problem (e.g. pole placement, optimal feedback, etc.) to the conventional decentralized control design problem. Design procedures for different types of controllers, such as periodic and sampled-data control laws, are also investigated. The notion of a quotient overlapping fixed mode (QOFM) is then introduced to determine whether the system can be stabilized by means of general (nonlinear and time-varying) structurally constrained controllers. It is shown that a system with no unstable QOFM can be stabilized by utilizing a finite-dimensional linear time-varying control law.

Bibliography

[1] S. H. Wang and E. J. Davison, "On the stabilization of decentralized control systems," *IEEE Transactions on Automatic Control*, vol. 18, no. 5, pp. 473-478, 1973.

[2] E. J. Davison and T. N. Chang, "Decentralized stabilization and pole assignment for general proper systems," *IEEE Transactions on Automatic Control*, vol. 35, no. 6, pp. 652-664, 1990.

[3] Z. Gong and M. Aldeen, "Stabilization of decentralized control systems," *Journal of Mathematical Systems, Estimation, and Control*, vol. 7, no. 1, pp. 1-16, 1997.

[4] A. I. Zecevic and D. D. Šiljak, "A new approach to control design with overlapping information structure constraint," *Automatica*, vol. 41, no. 2, pp. 265-272, 2005.

[5] L. Bakule, J. Rodellar, and J. M. Rossell, "Contractibility of dynamic LTI controllers using complementary matrices," *IEEE Transactions on Automatic Control*, vol. 48, no. 7, pp. 1269-1274, 2003.

[6] A. Iftar, "Overlapping decentralized dynamic optimal control," *International Journal of Control*. vol. 58, no. 1, pp. 187-209, 1993.

[7] A. Iftar, "Decentralized optimal control with overlapping decompositions," *IEEE International Conference on Systems Engineering*, pp. 299-302, 1991.

[8] D. D. Šiljak and A. I. Zecevic, "Control of large-scale systems: Beyond decentralized feedback," *Annual Reviews in Control*, vol. 29, no. 2, pp. 169-179, 2005.

[9] P. P. Groumpos, "Structural modelling and optimisation of large scale systems," *IEE Control Theory and Applications*, vol. 141, no. 1, pp. 1-11, 1994.

[10] A. P. Leros and P. P. Groumpos, "The time-invariant BAS decentralized large-scale linear regulator proble," *International Journal of Control*, vol. 46, no. 1, pp. 129-152, 1987.

[11] S. S. Stankovic and M. J. Stanojevic, and D. D. Šiljak, "Decentralized overlapping control of a platoon of vehicles," *IEEE Transactions on Control Systems Technology*, vol. 8, no. 5, pp. 816-832, 2000.

[12] Y. Ebihara and T. Hagiwara, "Structured controller synthesis using LMI and alternating projection method," *Proceedings of the 42nd IEEE Conference on Decision and Control*, pp. 5632-5637, 2003.

[13] J. Han and R. E. Skelton, "An LMI optimization approach for structured linear controllers," *Proceedings of the 42nd IEEE Conference on Decision and Control*, pp. 5143-5148, 2003.

[14] M. S. Ravi, J. Rosenthal, and X. A. Wang, "On decentralized dynamic pole placement and feedback stabilization," *IEEE Transactions on Automatic Control*, vol. 40, no. 9, pp. 1603-1614, 1995.

[15] D. D. Šiljak, Decentralized control of complex systems, Cambridge: Academic Press, 1991.

[16] K. D. Young,"On near optimal decentralized control," *Automatica*, vol. 21, no. 5, pp. 607-610, 1985.

[17] H. T. Toivoneh and P. M. Makila, "A descent anderson-moore algorithm for optimal decentralized control," *Automatica*, vol. 21, no. 6, pp. 743-744, 1985.

[18] M. Rossi and D. E. Miller,"Gain/phase margin improvement using static generalized sampled-data hold functions," *Systems & Control Letters*, vol. 37, no. 3, pp. 163-172, 1999.

[19] P. T. Kabamba, "Control of linear systems using generalized sampled-data hold functions," *IEEE Transactions on Automatic Control*, vol. 32, no. 9, pp. 772-783, 1987.

[20] J. L. Yanesi and A. G. Aghdam, "Optimal generalized sampled-data hold functions with a constrained structure," *Proceedings of the 2006 American Control Conference*, Minneapolis, Minnesota, 2006.

[21] A. G. Aghdam, "Decentralized control design using piecewise constant hold functions," *Proceedings of the 2006 American Control Conference*, Minneapolis, Minnesota, 2006.

[22] S. H. Wang, "Stabilization of decentralized control systems via time-varying controllers," *IEEE Transactions on Automatic Control*, vol. 27, no. 3, pp. 741-744, 1982.

[23] Ü. Özgüner and E. J. Davison, "Sampling and decentralized fixed modes," *Proceedings of the 1985 American Control Conference*, pp. 257-262, 1985.

[24] B. Anderson and J. Moore, "Time-varying feedback laws for decentralized control," *IEEE Transactions on Automatic Control*, vol. 26, no. 5, pp. 1133-1139, 1981.

[25] V. Pichai, M. E. Sezer, and D. D. Šiljak, "A graph-theoretic characterization of structurally fixed modes," *Automatica*, vol. 20, no. 2, pp. 247-250, 1984.

Chapter 9

Robust Stability Verification using Sum-of-Squares

9.1 Abstract

This paper deals with the robust stability of discrete-time LTI systems with parametric uncertainties belonging to a semi-algebraic set. It is asserted that the robust stability of any system over any semi-algebraic set (satisfying a mild condition) is equivalent to solvability of a semidefinite programming (SDP) problem, which can be handled using the available software tools. The particular case of a semi-algebraic set associated with a polytope is then investigated, and a computationally appealing method is proposed to attain the SDP problem by means of the sampling technique, introduced recently in the literature. Furthermore, it is shown that the current results encompass the ones presented in some of the recent works.

9.2 Introduction

Robust stability verification of a system subject to the parametric uncertainty has attracted many researchers in recent years [1-8]. The uncertainty is often assumed to belong to a polytope. So far, the most efficient technique in the literature to tackle this problem has been to check the existence of a proper Lyapunov function. The earlier works were based on seeking a constant Lyapunov function. While such approaches are appealing in terms of the computational requirements, they may arrive at very conservative solutions in general. It is shown in [9] that among numerous sorts of relations which can be considered for the Lyapunov function, it suffices to only consider polynomials. One of the recent works which leads to a simple condition is [4], where the Lyapunov function is implicitly assumed to be a polynomial of degree one. Moreover, an inequality is used in [4] to obtain the LMI conditions, which turns to an equality only when a parameter-dependent function is constant over the polytope. Therefore, this method proves to be very conservative in general, due to its restrictive formulation. Another approach is presented in [3], which leads to several inequalities obtained through a conservative procedure (this approach considers first-order Lyapunov functions).

In [8], a number of LMI conditions are attained which assure the robust stability of a system over an affine space. The Lyapunov function is assumed to belong to a class of matrix polynomials with any arbitrary degree, and it is shown that by increasing the degree of the Lyapunov function, the conditions obtained become less conservative. However, no bound on the degree of the corresponding Lyapunov function is attained.

It is shown in [2] that a continuous-time system is robustly stable over a polytope if and only if there exists a Lyapunov function in the form of a homogeneous polynomial with a specific bound on its degree such that it is positive definite over the polytope. A sufficient condition is obtained subsequently, by writing a nonnegative homogeneous matrix polynomial as a sum of some squared matrix polynomials. The degree of conservativeness of this work depends on the possibility of such a representation.

The work [1] seeks a Lyapunov function in the form of a homogeneous polynomial as well (like the work [2]). It is shown that as the degree of the Lyapunov function increases, the conservativeness of the resultant LMI conditions reduces. However, no convergence proof is provided in [1].

The present work deals with the robust stability of discrete-time systems over a semi-algebraic set, particularly a polytope. It is shown that the system is robustly stable over any semi-algebraic set satisfying a mild condition if and only if there exist a number of matrix polynomials with certain bounds on their degrees, satisfying a specific relation along with the Lyapunov function. This relation can be simply converted to a semidefinite programming (SDP) problem, which presents a necessary and sufficient condition for robust stability in the form of nonparametric SDP or LMI. Moreover, the special case of polytopic uncertainty is studied and it is shown how the parametric equation obtained can be solved numerically by means of sampling technique, introduced in the literature recently [11].

It is finally shown that the results presented here encompass the ones given in [1] and [2] for discrete-time systems.

9.3 Problem formulation

Consider an uncertain discrete-time system $\mathscr{S}$ of order ν with the following state equation:

$$x(\kappa+1) = A(\alpha)x(\kappa), \quad \kappa = 0,1,2,... \tag{9.1}$$

where the vector $\alpha = \begin{bmatrix} \alpha_1 & \alpha_2 & \cdots & \alpha_m \end{bmatrix}$ is used to represent the perturbation of the system matrix $A(\alpha) \in \Re^{\nu\times\nu}$. Suppose that $A(\alpha)$ is a matrix polynomial of degree σ and that it can be expressed in terms of the known matrices $A_1,...,A_n$ and polynomials $g_1(\alpha), g_2(\alpha),..., g_n(\alpha)$ as:

$$A(\alpha) = g_1(\alpha)A_1 + g_2(\alpha)A_2 + \cdots + g_n(\alpha)A_n \tag{9.2}$$

Given the polynomials $q_1(\alpha),...,q_\eta(\alpha)$, define the semi-algebraic set $\mathscr{D}$ as follows:

$$\mathscr{D} := \left\{\alpha \,\middle|\, q_1(\alpha) \geq 0,...,q_\eta(\alpha) \geq 0\right\} \tag{9.3}$$

and assume that the region $\mathscr{D}$ is the closure of some open connected set. It is desired now to find out whether the system $\mathscr{S}$ is robustly stable for any α belonging to the semi-algebraic set $\mathscr{D}$ defined above. The following notations and definitions will prove convenient in presenting the main results.

Notation 1: Bold symbols are used throughout the paper to denote the vector of variables of any multivariate scalar/matrix polynomial.

Notation 2: For any vector $\omega = \begin{bmatrix} \omega_1 & \omega_2 & \cdots & \omega_l \end{bmatrix}$, define ω^2 as $\begin{bmatrix} \omega_1^2 & \omega_2^2 & \cdots & \omega_l^2 \end{bmatrix}$.

Notation 3: Let the following symbols be introduced:

- Given a symmetric matrix M, denote its maximum and minimum eigenvalues with $\bar{\lambda}(M)$ and $\underline{\lambda}(M)$, respectively. Besides, denote the spectral norm and the Frobenius norm of the matrix M with $||M||_2$ and $||M||_F$, respectively.

- Given a matrix polynomial $C(\alpha)$, denote the degree of $C(\alpha)$ with $\deg(C)$.

Definition 1: Given a matrix polynomial $C(\alpha)$, let each term of this polynomial be represented as $\Gamma_g \alpha_1^{g_1}\alpha_2^{g_2}...\alpha_\mu^{g_\mu}$. Compute the quantity $\frac{g_1!g_2!...g_\mu!||\Gamma_g||_2}{(g_1+g_2+\cdots+g_\mu)!}$ for all different terms of this polynomial and denote the maximum one with $||C(\alpha)||_s$ (see [12]).

Definition 2: A matrix polynomial $C(\omega)$ is defined to be a sum-of-squares (SOS) if there exists a matrix polynomial $E(\omega)$ such that $C(\omega) = E^T(\omega)E(\omega)$.

9.4 A necessary and sufficient condition for robust stability

To develop the main results of this work, assume that there exist a scalar r and SOS polynomials $h_0(\omega), h_1(\omega), \ldots, h_k(\omega)$ such that $r^2 - \omega\omega^T = h_0(\omega) + \sum_{i=1}^{\eta} h_i(\omega)q_i(\omega)$ (note that using a proper SOS toolbox, one can check if this mild condition is satisfied). Furthermore, assume that $\mathscr{D}$ is the closure of some open connected set.

Lemma 1: The system $\mathscr{S}$ is robustly stable in the domain $\mathscr{D}$ if and only if there exists a Lyapunov matrix polynomial $P(\alpha)$ of degree at most $\zeta_1 := (\nu^2 + \nu - 2)\sigma$ such that $\Phi(\alpha)$ is positive definite for any α belonging to the region $\mathscr{D}$, where:

$$\Phi(\alpha) := \begin{bmatrix} P(\alpha) & A^T(\alpha)P(\alpha) \\ P(\alpha)A(\alpha) & P(\alpha) \end{bmatrix} \tag{9.4}$$

Proof: It is well-known that the system $\mathscr{S}$ is robustly stable in the domain $\mathscr{D}$ if and only if there exists a Lyapunov function $H(\alpha)$ which is positive definite over $\mathscr{D}$ and satisfies the equation:

$$H(\alpha) - A^T(\alpha)H(\alpha)A(\alpha) = I \tag{9.5}$$

By exploiting Cramer's rule and in line with the proof of Lemma 1 in [2], it can be shown that $H(\alpha)$ can be written as $\frac{P(\alpha)}{h(\alpha)}$, where $P(\alpha)$ and $h(\alpha)$ are coprime matrix and scalar polynomials of respective maximum degrees ζ_1 and $\zeta + 2\sigma$. For a robustly stable system, both of the polynomials $P(\alpha)$ and $h(\alpha)$ are positive definite over the region $\mathscr{D}$ (see Corollary 2 in [13]). It follows from this observation along with the Schur complement formula that the matrix $\Phi(\alpha)$ constructed in terms of $P(\alpha)$ must be positive definite over $\mathscr{D}$, for the system $\mathscr{S}$ to be robustly stable. ■

Theorem 1 *The system $\mathscr{S}$ is robustly stable in the region $\mathscr{D}$ if and only if there exist a positive scalar ε, a matrix polynomial $P(\alpha)$ of degree at most ζ_1 and SOS matrix polynomials $Q_0(\alpha), \ldots, Q_\eta(\alpha)$ such that:*

$$\Phi(\alpha) = Q_0(\alpha) + \sum_{i=1}^{\eta} q_i(\alpha)Q_i(\alpha) + \varepsilon I, \qquad \forall \alpha \in \Re^m \tag{9.6}$$

Proof: The proof follows immediately from Theorem 2 in [14] and Lemma 1 given above. ■

Solvability of the matrix equation (9.6) can be checked using the available software tools (e.g. YALMIP and SOSTOOLS [15; 16]) provided some bounds on the degrees of the polynomials $Q_0(\alpha)$, ..., $Q_\eta(\alpha)$ are known. This issue will be discussed later in Subsection B.

9.4.1 Polytopic uncertainty

Assume that $m = n$ and $g_i(\alpha) = \alpha_i$, for all $i = 1, 2, ..., n$. Consider the polytopic region:

$$\mathscr{P} := \left\{ \alpha \mid 0 \leq \alpha_1, ..., \alpha_n \leq 1, \sum_{i=1}^{n} \alpha_i = 1 \right\} \tag{9.7}$$

As a special case, robust stability analysis for the system $\mathscr{S}$ over the region $\mathscr{P}$ is of particular interest in the literature (note that the region $\mathscr{P}$ is a semi-algebraic set). Define now the following matrices:

$$\bar{\Phi}(\alpha) := \begin{bmatrix} P(\alpha)\left(\sum_{i=1}^{n} \alpha_i\right) & A^T(\alpha)P(\alpha) \\ P(\alpha)A(\alpha) & P(\alpha)\left(\sum_{i=1}^{n} \alpha_i\right) \end{bmatrix}, \quad \tilde{\Phi}(\alpha) := \bar{\Phi}(\alpha)\left(\sum_{i=1}^{n} \alpha_i\right) \tag{9.8}$$

Since the term $\sum_{i=1}^{n} \alpha_i$ is equal to 1 over the polytope, one can express Lemma 1 in terms of the function $\bar{\Phi}(\alpha)$ or $\tilde{\Phi}(\alpha)$, instead of $\Phi(\alpha)$ in (9.4) (by considering $\mathscr{D} = \mathscr{P}$).

Theorem 2 *The system $\mathscr{S}$ is robustly stable in the domain $\mathscr{P}$ if and only if there exist a scalar $\varepsilon > 0$, a homogeneous matrix polynomial $P(\alpha)$, a matrix polynomial $Q_1(\omega)$, and an SOS matrix polynomial $Q_2(\omega)$, where $\omega = \begin{bmatrix} \omega_1 & \omega_2 & \cdots & \omega_n \end{bmatrix}$, such that any one of the following three equations (whose feasibilities are equivalent indeed) holds:*

$$\Phi\left(\omega^2\right) = \left(1 - \omega\omega^T\right)Q_1(\omega) + Q_2(\omega) + \varepsilon I_{2v} \tag{9.9a}$$

$$\bar{\Phi}\left(\omega^2\right) = \left(1 - \omega\omega^T\right)Q_1(\omega) + Q_2(\omega) + \varepsilon I_{2v} \tag{9.9b}$$

$$\tilde{\Phi}\left(\omega^2\right) = \left(1 - \omega\omega^T\right)Q_1(\omega) + Q_2(\omega) + \varepsilon I_{2v} \tag{9.9c}$$

Proof: The proof follows from Theorem 1, and on noting that:

i) The homogeneousness of $P(\alpha)$ is proved in [2].

ii) By changing the variable α to ω^2, the region $\mathscr{P}$ will be mapped into the domain $\mathscr{D} := \{\omega \mid \omega\omega^T - 1 \geq 0, 1 - \omega\omega^T \geq 0\}$, which has the form defined by (9.3). ■

It will later be shown (in Section IV) that the formulas (9.9b) and (9.9c) encompass the formulations given in [1] and [2], respectively.

Corollary 1: The system $\mathscr{S}$ is robustly stable over the polytope $\mathscr{P}$ if there exist symmetric matrices $P_1, P_2, ..., P_n$ and symmetric positive definite matrices $Z_{ij},\ i, j \in \{1, 2, ..., n\},\ i < j$, such that the symmetric matrix U given by the following diagonal block entries:

$$U_{ii} = \begin{bmatrix} P_i & A_i'P_i \\ P_iA_i & P_i \end{bmatrix}, \quad \forall\, i \in \{1, 2, ..., n\} \tag{9.10}$$

and the off-diagonal block entries:

$$U_{ij} = \frac{1}{2} \begin{bmatrix} P_i + P_j & A_i'P_j + A_jP_i \\ P_iA_j + P_jA_i & P_i + P_j \end{bmatrix} - Z_{ij}, \quad \forall\, i, j \in \{1, 2, ..., n\}, \quad i < j \tag{9.11}$$

is positive definite.

Proof: The proof is an immediate consequence of the following choices of matrices:

$$P(\alpha) = \sum_{i=1}^{n} P_i\alpha_i, \quad Q_1(\omega) = -I_{2v} + \sum_{i=1}^{n} F_i\omega_i^2, \quad Q_2(\omega) = \sum_{i,j=1}^{n} F_{ij}\omega_i^2\omega_j^2 \tag{9.12}$$

in the equation (9.9a). The details are omitted here. ■

The SOS equations given in (9.9) can be converted to SDP problems, in which the polynomial $Q_1(\omega)$ is eliminated via the sampling technique introduced in [11; 10]. This conversion will be explained now for the equation (9.9a). Given the maximum degrees of $P(\alpha)$ and $Q_2(\omega)$, assume that the solvability of the equation (9.9a) is desired to be checked. The following algorithm is proposed for this purpose.

Algorithm 1:

Step 1. Find the number of unknown coefficients of $P(\alpha)$ and $Q_2(\omega)$ and denote it with ζ (which depends on $\deg(P)$ and $\deg(Q_2)$).

Step 2. Choose ζ generic points in the closed unit ball of dimension $n-1$, and denote their coordinates with $\bar{\lambda}_i = \begin{bmatrix} \lambda_{i_1} & \lambda_{i_2} & \cdots & \lambda_{i_{n-1}} \end{bmatrix}$, for $i = 1, 2, ..., \zeta$.

Step 3. Define $\lambda_{i_n} = \sqrt{\lambda_{i_1}^2 + \cdots + \lambda_{i_{n-1}}^2}$ and $\lambda_i = \begin{bmatrix} \lambda_{i_1} & \lambda_{i_2} & \cdots & \lambda_{i_n} \end{bmatrix}$, for $i = 1, 2, ..., \zeta$.

Step 4. The robust stability of the system $\mathscr{S}$ over the polytope $\mathscr{P}$ can be determined by the polynomials $P(\alpha)$ and $Q_2(\omega)$ of pre-specified degrees if and only if the following SDP problem is solvable:

$$\begin{bmatrix} P\left(\lambda_i^2\right) & A^T\left(\lambda_i^2\right)P\left(\lambda_i^2\right) \\ P\left(\lambda_i^2\right)A\left(\lambda_i^2\right) & P\left(\lambda_i^2\right) \end{bmatrix} - Q_2(\lambda_i) - I_{2v} = 0, \;\; i = 1, 2, ..., \zeta \tag{9.13}$$

The details of the above procedure have been investigated thoroughly in [17].

Remark 1: The term *generic* in Algorithm 1 signifies that the point $\begin{bmatrix} \bar{\lambda}_1 & \cdots & \bar{\lambda}_\zeta \end{bmatrix}$ must not lie on a particular hypersurface in the $\zeta.(n-1)$ dimensional space. One can refer to [17] for more details.

Remark 2: Algorithm 1, if its first step is excluded, can be inferred from the work [11]. In fact, the number ζ proposed in [11] also takes the coefficients of $Q_1(\omega)$ into account, and is thus greater than the one given in Step 1 of Algorithm 1. This in turn leads to a more sophisticated SDP problem.

Remark 3: Choosing ζ points at random in Step 2 of Algorithm 1 may result in an ill-conditioned problem [18; 19], in light of the potential numerical difficulty of extrapolating a polynomial from its sample points. Nevertheless, there are some results in the literature to remedy this problem [20]. For instance, one can choose a set of ζ generic *complex* points (instead of ζ real points in the closed unit ball) in order to obtain a well-conditioned interpolation problem [18].

9.4.2 Bounds on the degrees of the polynomials $Q_0(\alpha), ..., Q_\eta(\alpha)$

Similar to the exiting works, it is not normally known *a priori* for what degrees of the polynomials $Q_0(\alpha), ..., Q_\eta(\alpha)$ the feasibility of the SOS formula given in Theorem 1 (or Theorem 2, as a particular form of it) needs to be checked. In this case, an infinite hierarchy of SOS problems should be constructed by repeatedly increasing the degrees of the relevant polynomials in Theorem 1 towards infinity. Now, one should check the feasibility of all the SOS problems in this hierarchy successively, until one of them turns out to be feasible. It is to be noted, however, that in order to prove the system is not robustly stable, infeasibility of infinitely many SOS problems needs to be verified. Note also that all the SOS approaches surveyed in this work encounter said shortcoming. To alleviate this issue, it is desirable to know some bounds on the degrees of the polynomials $Q_0(\alpha), ..., Q_\eta(\alpha)$. Since obtaining these bounds may not be realistic in general, it is preferred instead to provide an algorithm that although the number of its iterations is initially unknown, it is known at any iteration whether the algorithm should proceed to the next iteration or it should halt.

Let ζ_2 denote any upper bound on the supremum of the polynomial $g_1(\alpha)^2 + \cdots + g_n(\alpha)^2$ over the compact region $\mathscr{D}$ (the bound ζ_2 can be obtained by employing the existing methods; e.g., the one in [13]). Furthermore, define $\zeta_3 := \|A_1\|_F^2 + \cdots + \|A_n\|_F^2$. Consider now $P(\alpha)$ and $h(\alpha)$ introduced in Lemma 1. It can be straightforwardly verified that such polynomials are not unique, unless the conditions $\|h(\alpha)\|_s = 1$ and $h(\alpha_o) \geq 0$ are imposed, where α_o denotes an arbitrary pre-determined point in $\mathscr{D}$. Unlike $P(\alpha)$, the coefficients of $h(\alpha)$ can be easily found by solving a set of linear numerical equations, as discussed in [21]. Thus, assume henceforth that $h(\alpha)$ is known.

Theorem 3 *Consider an arbitrary point α_* in the domain $\mathscr{D}$. Assume that the system $\mathscr{S}$ is robustly stable over $\mathscr{D}$ with a Lyapunov polynomial $P(\alpha)$ satisfying the equation (9.5). Then, the following inequality holds:*

$$\frac{\zeta_4}{\sqrt{3\nu + 2\zeta_2\zeta_3 + 2\zeta_2^2\zeta_3^2}} < \underline{\lambda}(\Phi(\alpha_*)) \tag{9.14}$$

where ζ_4 is any strict lower bound on the minimum of $h(\alpha)$ in the region $\mathscr{D}$.

Sketch of proof: It is straightforward to show that (using Lemma 2.2 in [22]):

$$\begin{aligned}\underline{\lambda}\left(\begin{bmatrix} I \times h(\alpha_*) & 0 \\ 0 & P(\alpha_*) \end{bmatrix}\right) &= \underline{\lambda}\left(\begin{bmatrix} I & -A^T(\alpha_*) \\ 0 & P(\alpha_*) \end{bmatrix} \Phi(\alpha_*) \begin{bmatrix} I & 0 \\ -A(\alpha_*) & P^{-1}(\alpha_*) \end{bmatrix}\right) \\ &\leq \bar{\lambda}\left(\begin{bmatrix} I & 0 \\ -A(\alpha_*) & P^{-1}(\alpha_*) \end{bmatrix} \begin{bmatrix} I & -A^T(\alpha_*) \\ 0 & P(\alpha_*) \end{bmatrix}\right) \times \underline{\lambda}(\Phi(\alpha_*))\end{aligned} \tag{9.15}$$

Moreover:

$$\begin{aligned}\bar{\lambda}\left(\begin{bmatrix} I & 0 \\ -A(\alpha_*) & P^{-1}(\alpha_*) \end{bmatrix} \begin{bmatrix} I & -A^T(\alpha_*) \\ 0 & P(\alpha_*) \end{bmatrix}\right) &\leq \left\| \begin{bmatrix} I & -A^T(\alpha_*) \\ -A(\alpha_*) & I + A^T(\alpha_*)A(\alpha_*) \end{bmatrix} \right\|_F \\ &\leq \sqrt{\nu + 2\|A(\alpha_*)\|_F^2 + \left(\sqrt{\nu} + \|A^T(\alpha_*)A(\alpha_*)\|\right)^2} \leq \sqrt{3\nu + 2\|A(\alpha_*)\|_F^2 + 2\|A(\alpha_*)\|_F^4}\end{aligned} \tag{9.16}$$

Furthermore, the Cauchy-Schwarz inequality yields that:

$$\|A(\alpha_*)\|_F^2 \leq \left(g_1(\alpha_*)^2 + \cdots + g_n(\alpha_*)^2\right)\left(\|A_1\|_F^2 + \cdots + \|A_n\|_F^2\right) = \zeta_2\zeta_3 \tag{9.17}$$

The proof follows from the equations (9.15), (9.16), (9.17), and on noting that all the eigenvalues of the matrix $P(\alpha_*)$ are greater than ζ_4 due to the equation (9.5). ■

Lemma 2: For every matrix polynomials $M_1(\alpha)$ and $M_2(\alpha)$, the relations given below hold:

$$\|M_1(\alpha)+M_2(\alpha)\|_s \leq \|M_1(\alpha)\|_s + \|M_2(\alpha)\|_s \tag{9.18a}$$

$$\|M_1(\alpha)M_2(\alpha)\|_s \leq (\deg(M_1)+1)(\deg(M_2)+1)\|M_1(\alpha)\|_s\|M_2(\alpha)\|_s \tag{9.18b}$$

Proof: The correctness of the inequality (9.18a) follows directly from Definition 1. Furthermore, the inequality (9.18b) has been proved in Proposition 14 of [12] for the scalar polynomials, and the results can be easily extended to the matrix polynomials. ■

Theorem 4 *Assume that the system $\mathscr{S}(\alpha)$ is robustly stable over $\mathscr{D}$, and that $\|A(0)\|_2 < \frac{1}{\sigma+1}$. Choose a number $\gamma \geq 1$ such that $\|A(\frac{\alpha}{\gamma})\|_s$ is less than $\frac{1}{\sigma+1}$ (due to the above assumption, such γ exits). The inequality $\|\Phi(\alpha)\|_s \leq \zeta_5$ holds, where:*

$$\zeta_5 := \frac{\gamma^{\zeta_1}((\zeta_1+1)(\sigma+1)\|A(\alpha)\|_s+1)(\zeta_1+2\sigma+1)\|h(\frac{\alpha}{\gamma})\|_s}{1-(\sigma+1)^2\|A(\frac{\alpha}{\gamma})\|_s^2} \tag{9.19}$$

Sketch of proof: It can be concluded from the relation $P(\alpha)-A^T(\alpha)P(\alpha)A(\alpha) = h(\alpha)\times I$ that:

$$P\left(\frac{\alpha}{\gamma}\right) = \sum_{i=0}^{\infty} A^T\left(\frac{\alpha}{\gamma}\right)^i h\left(\frac{\alpha}{\gamma}\right) A\left(\frac{\alpha}{\gamma}\right)^i, \quad \forall \alpha \in \gamma\mathscr{D} \tag{9.20}$$

Thus, it follows from Lemma 2 and the assumption $\|A(\frac{\alpha}{\gamma})\|_s < \frac{1}{\sigma+1}$ that:

$$\left\|P\left(\frac{\alpha}{\gamma}\right)\right\|_s \leq (\deg(h)+1)\left\|h\left(\frac{\alpha}{\gamma}\right)\right\|_s \sum_{i=0}^{\infty}\left((\sigma+1)^{2i}\left\|A\left(\frac{\alpha}{\gamma}\right)\right\|_s^{2i}\right) = \frac{(\deg(h)+1)\|h(\frac{\alpha}{\gamma})\|_s}{1-(\sigma+1)^2\|A(\frac{\alpha}{\gamma})\|_s^2} \tag{9.21}$$

On the other hand, it is straightforward to show that:

$$\|\Phi(\alpha)\|_s \leq \|P(\alpha)\|_s + (\deg(P)+1)(\sigma+1)\|P(\alpha)\|_s\|A(\alpha)\|_s \tag{9.22}$$

The proof is completed by combining the inequalities (9.21) and (9.22), and taking the relation $\frac{\|P(\alpha)\|_s}{\gamma^{\deg(P)}} \leq \left\|P\left(\frac{\alpha}{\gamma}\right)\right\|_s$ into consideration (note that $\gamma \geq 1$ and $\deg(P) = \deg(h) - 2\sigma = \zeta_1$). ■

It is worth mentioning that the assumption made on the 2-norm of $A(0)$ in Theorem 4 is automatically satisfied if the matrix polynomial $A(\alpha)$ has no constant term (for instance, this is the case for the polytopic uncertainty). In what follows, a bound on the size of the SOS problem to be solved in the polytopic case will be obtained, which can be extended to the general case of non-polytopic uncertainty in a similar way.

Theorem 5 *Choose a real number $\gamma \geq 1$ with the property that $\|A(\frac{\alpha}{\gamma})\|_s < \frac{1}{2}$, and assume that ζ_4 (defined in Theorem 3) is strictly positive. For checking the robust stability of the system $\mathscr{S}$ over $\mathscr{P}$ using the equation (9.9b) without introducing any conservativeness, the degrees of the polynomials $\Phi_1(\omega)$ and $\Phi_2(\omega)$ need not exceed $\Gamma - 2$ and Γ respectively, where:*

$$\Gamma := \frac{2\gamma^{\zeta_1}\zeta_1(\zeta_1+3)(\zeta_1+1)^2(\|A(\alpha)\|_s+1)\|h(\frac{\alpha}{\gamma})\|_s\sqrt{3\nu+2\zeta_2\zeta_3+2\zeta_2^2\zeta_3^2}}{\zeta_4\left(1-4\|A(\frac{\alpha}{\gamma})\|_s^2\right)} \tag{9.23}$$

Proof: Assume that the system $\mathscr{S}$ is robustly stable with a Lyapunov function $P(\alpha)$ satisfying the equation (9.5). Define now ε as a positive number sufficiently smaller than the absolute difference between the two sides of the inequality (9.14). According to Theorem 3 and due to the relation $\min_{\alpha\in\mathscr{P}} \underline{\lambda}(\bar{\Phi}(\alpha)) = \min_{\alpha\in\mathscr{P}} \underline{\lambda}(\Phi(\alpha))$, the matrix $\bar{\Phi}(\alpha) - \varepsilon I$ is nonnegative definite over the polytope $\mathscr{P}$. Hence, it can be concluded from Theorem 3 of [14] that all the coefficients of $(\alpha_1 + \alpha_2 + \cdots + \alpha_m)^\beta(\bar{\Phi}(\alpha) - \varepsilon I)$ are nonnegative definite, where $\beta = \frac{\deg(\bar{\Phi})(\deg(\bar{\Phi})-1)\|\bar{\Phi}(\alpha)-\varepsilon\|_s}{2\min_{\alpha\in\mathscr{P}}\underline{\lambda}(\bar{\Phi}(\alpha)-\varepsilon)} - \deg(\bar{\Phi})$. This implies that $(\omega\omega^T)^\beta(\bar{\Phi}(\omega^2) - \varepsilon I)$ has only nonnegative definite coefficients. Extending Proposition 2 of [23] to the matrix polynomial case, it can subsequently be deduced from the latter result that there exist a matrix polynomial $Q_1(\omega)$ and an SOS matrix polynomial $Q_2(\omega)$ (with all monomials having even degrees) satisfying the equation (9.9b), where the degrees of $Q_1(\omega)$ and $Q_2(\omega)$ do not exceed $2\beta + 2\deg(\bar{\Phi})$ and $2\beta + 2\deg(\bar{\Phi}) - 2$, respectively. Now, the proof follows by combining the results given below:

- $\deg(\bar{\Phi})$ is less than or equal to $\zeta_1 + 1$ (see Lemma 1).
- A lower bound on $\min_{\alpha\in\mathscr{P}} \underline{\lambda}(\bar{\Phi}(\alpha) - \varepsilon)$ is given in the left side of the inequality (9.14).
- Since ε is sufficiently small, $\|\bar{\Phi}(\alpha) - \varepsilon\|_s$ is sufficiently close to $\|\bar{\Phi}(\alpha)\|_s$, and an upper bound for it can be obtained as follows (similarly to the one given in Theorem 4):

$$\|\bar{\Phi}(\alpha)\|_s \leq \frac{\gamma^{\zeta_1}((\zeta_1+1)(1+1)\|A(\alpha)\|_s + (\zeta_1+1)(1+1))(\zeta_1+3)\|h(\frac{\alpha}{\gamma})\|_s}{1-2^2\|A(\frac{\alpha}{\gamma})\|_s^2} \tag{9.24}$$

(note that σ is equal to 1 in the polytopic case). ■

Theorem 5 states that when a positive *strict* lower bound ζ_4 on the polynomial $h(\alpha)$ over the polytope $\mathscr{P}$ is provided, the size of the SOS problem required to check the robust stability of the system can be computed systematically. Nevertheless, the question arises whether such a lower bound can be obtained (when it exists). To answer this question, one can take advantage of the SOS method proposed in [24] for obtaining *a non-decreasing sequence converging to the minimum of* $h(\alpha)$ *over the region* $\mathscr{P}$, along with the one in [25] for obtaining *a non-increasing sequence converging to the minimum of* $h(\alpha)$ (note that $h(\alpha)$ is homogeneous). One can employ these two methods *alternately* to obtain the elements of these two sequences successively. One of the following three scenarios will certainly occur:

i) At some iteration, an element of the non-decreasing sequence becomes positive. Thus, ζ_4 can be considered as any positive number less than this value.

ii) At some iteration, an element of the non-increasing sequence becomes negative. Hence, it follows from Corollary 1 in [13] that the Lyapunov function $H(\alpha)$ (introduced in Lemma 1) is not nonnegative definite over the whole space; consequently, it results from the homogeneousness property of $H(\alpha)$ that it is not nonnegative definite over the polytope $\mathscr{P}$ either. This means that the system is not robustly stable.

iii) Both of the sequences are constantly converging to zero. In order to prevent infinite iterations in this case and to ensure that the corresponding minimum is not equal to zero, one can use the result of [26] for checking whether $h(\alpha)$ and the polytope have any common zeros. If they do, the system is not robustly stable (see the proof of Lemma 1); otherwise continue obtaining the elements of the two sequences until either case (i) or case (ii) occurs.

Once a positive value for ζ_4 is attained, the size of the main SOS problem to be solved can be determined. Although obtaining ζ_4 requires a finite, nevertheless indefinite number of iterations, it is known at each step whether the algorithm should proceed or stop. Note that the algorithm terminates due to either a positive ζ_4 being obtained, or the system not being robustly stable.

9.5 Generality of the proposed results

9.5.1 Comparison with [1]

The work [1] presents a sufficient condition for the polytopic robust stability based on the fact that the system is robustly stable over the polytope if there exists a homogeneous polynomial $P(\alpha)$ for which the coefficients of $\bar{\Phi}(\alpha)$ are all positive definite. Since the matrix $\bar{\Phi}(\omega^2)$ will be SOS in this case, the robust stability condition in [1] can be considered as a special case of the result provided in this work by setting $Q_1(\omega)$ in Theorem 3 to zero.

9.5.2 Comparison with [2]

According to [2], the system $\mathscr{S}$ is robustly stable over the polytope $\mathscr{P}$ if there exists a homogeneous Lyapunov function $P(\omega)$ with a bound on its degree such that:

$$P(\omega^2) > 0, \quad P(\omega^2)\left(\omega\omega^T\right)^2 - A(\omega^2)^T P(\omega^2) A(\omega^2) > 0, \qquad \forall \omega \in \Re^m \tag{9.25}$$

The above matrices are then assumed to be SOS, rather than just positive, in order to simplify the problem. A method is then proposed to convert the corresponding problem to an SDP one. Assume now that the system $\mathscr{S}$ satisfies the robust stability condition of [2] over the polytope $\mathscr{P}$. This means that there exist two matrix polynomials $\Gamma_1(\omega)$ and $\Gamma_2(\omega)$ such that:

$$P(\omega^2) = \Gamma_1(\omega)\Gamma_1(\omega)^T, \quad P(\omega^2)\left(\omega\omega^T\right)^2 - A(\omega^2)^T P(\omega^2) A(\omega^2) = \Gamma_2(\omega)\Gamma_2(\omega)^T \tag{9.26}$$

It can be easily verified that:

$$\tilde{\Phi}(\omega^2) = \begin{bmatrix} \Gamma_2(\omega) & -A^T(\omega^2)\Gamma_1(\omega) \\ 0 & \Gamma_1(\omega)\omega\omega^T \end{bmatrix} \begin{bmatrix} \Gamma_2(\omega) & -A^T(\omega^2)\Gamma_1(\omega) \\ 0 & \Gamma_1(\omega)\omega\omega^T \end{bmatrix}^T \tag{9.27}$$

which means that $\tilde{\Phi}(\omega^2)$ is SOS. Now, one can choose $Q_1(\omega)$ in Theorem 3 to be 0, yielding that the condition given in [2] is encompassed by the one presented here.

9.6 Numerical example

The following example is handled by utilizing the software YALMIP and the solver SeDuMi. Moreover, a 2GHz processor is used to carry out the numerical computations, for the sake of consistency in comparing the corresponding processing times.

Assume that A_1 and A_2 are given by:

$$A_1 = T_1 \times \text{diag}([-1\ \ 0.4\ \ -0.97\ \ 0.99]) \times T_1^{-1},\ A_2 = T_2 \times \text{diag}([1\ \ -0.3\ \ 0.97\ \ -0.9]) \times T_2^{-1} \quad (9.28)$$

where:

$$T_1 = \begin{bmatrix} 1 & 2 & 0.3 & 0.4 \\ 3 & 2 & 1 & 0.4 \\ -1 & 2 & 3 & -1 \\ 0 & 0 & 0.6 & 1 \end{bmatrix}, \quad T_2 = \begin{bmatrix} 0.3 & 2 & 1 & 0.4 \\ -1 & -2 & -3 & 0 \\ -3 & 2 & -1 & 0 \\ -1 & -3 & -2 & 1 \end{bmatrix} \quad (9.29)$$

It is desired to obtain the maximum value of μ, for which the matrix $A(\alpha) = \mu(\alpha_1 A_1 + \alpha_2 A_2)$ is Schur for all $\alpha_1, \alpha_2 \in \mathscr{P}$. Since many of the recent works consider only first-order Lyapunov functions, let this type of Lyapunov function be sought first. Some of the recent results will now be employed to find μ.

1. The work [1] translates the robust stability verification to a LMI problem with two symmetric matrices P_1 and P_2 subject to the following inequalities.

$$\begin{bmatrix} P_1 & A_1^T P_1 \\ P_1 A_1 & P_1 \end{bmatrix} > 0, \begin{bmatrix} P_1 + P_2 & A_1^T P_2 + A_2^T P_1 \\ P_1 A_2 + P_2 A_1 & P_1 + P_2 \end{bmatrix} > 0, \begin{bmatrix} P_2 & A_2^T P_2 \\ P_2 A_2 & P_2 \end{bmatrix} > 0 \quad (9.30)$$

The value of μ obtained by solving the above LMI problem is 0.623, and the time it takes to attain this result is 0.219sec.

2. According to [2], the matrix $A(\alpha)$ is Schur over the polytope $\mathscr{P}$ if there exist three symmetric positive definite matrices P_1, P_2 and U, and two symmetric matrices Z_1 and Z_2, satisfying the following constraints:

$$\begin{aligned} &U_{11} = P_1 - A_1 P_1 A_1, \quad U_{22} = P_2 - A_2 P_2 A_2,\ U_{12} + U_{12}^T + U_{34} + U_{34}^T = 0, \\ &U_{13} = \frac{2P_1 + P_2 - A_1' P_1 A_2 - A_2' P_1 A_1 - A_1' P_2 A_1 - Z_2}{2},\ U_{23} + U_{23}^T = 0,\ U_{44} = Z_2, \\ &U_{24} = \frac{2P_2 + P_1 - A_2' P_2 A_1 - A_2' P_1 A_2 - A_1' P_2 A_2 - Z_1}{2},\ U_{14} + U_{14}^T = 0,\ U_{33} = Z_1 \end{aligned} \quad (9.31)$$

where U_{ij} denotes the (i, j) block entry of the matrix U for any $i, j \in \{1, 2, 3, 4\}$. Solving this LMI problem results in $\mu = 0.900$ with the processing time of 0.341sec.

3. According to [3], the matrix $A(\alpha)$ is Schur over $\mathscr{P}$ if there exist six symmetric matrices $P_1, P_2, Z_1, Z_2, Z_3, Z_4$ and two matrices Z_5 and Z_6 such that:

$$\begin{aligned} &A_1^T P_1 A_2 + A_2^T P_1 A_1 + A_1^T P_2 A_1 - 2P_1 - P_2 \leq Z_3 + Z_5 + Z_5^T, \\ &A_2^T P_2 A_1 + A_1^T P_2 A_2 + A_2^T P_1 A_2 - 2P_2 - P_1 \leq Z_2 + Z_6 + Z_6^T, \\ &\begin{bmatrix} Z_1 & Z_5 \\ Z_5^T & Z_2 \end{bmatrix} \leq 0, \quad \begin{bmatrix} Z_3 & Z_6 \\ Z_6^T & Z_4 \end{bmatrix} \leq 0, \quad A_1^T P_1 A_1 - P_1 < Z_1, \quad A_2^T P_2 A_2 - P_2 < Z_4 \end{aligned} \tag{9.32}$$

After elapsing 0.265sec, the value $\mu = 0.900$ is obtained in this case.

4. The work [4] states that the system is robustly stable if there exist a matrix G and two symmetric matrices P_1 and P_2 with the following properties:

$$\begin{bmatrix} P_1 & A_1^T G^T \\ G A_1 & G + G^T - P_1 \end{bmatrix} > 0, \quad \begin{bmatrix} P_2 & A_2^T G^T \\ G A_2 & G + G^T - P_2 \end{bmatrix} > 0 \tag{9.33}$$

The value $\mu = 0.633$ is obtained after 0.204sec.

5. Corollary 1 of the present paper results in $\mu = 0.720$ with the processing time of 0.166sec.

6. Theorem 3 (by considering $Q_1(\omega)$ as a first-order polynomial in the equation (9.9c)) leads to $\mu = 0.900$ in 0.285sec.

Comparing the results obtained, it can be concluded that Corollary 1 consumes the least time, while it leads to a less conservative result compared to [1] and [4]. Moreover, while the works [2], [3], and Theorem 3 all arrive at $\mu = 0.900$, the works [3] and [2] are the most and the least computationally efficient methods (time-wise), respectively. Although this shows an advantage of [3], the most important weak point of this work is its inability to obtain the exact value of μ, which is 0.927 in the above example. This is due to the restriction of considering first-order Lyapunov polynomials only. One can verify that using a second-order Lyapunov function in Example 1, the exact value given above will be achieved for μ using both [2] and Theorem 3 of the present work. However, the latter work elapses 0.328sec to arrive at this value; 17% faster than the former one.

9.7 Conclusions

In this paper, the robust stability of discrete-time LTI systems with uncertainties belonging to a semi-algebraic set satisfying a mild compactness condition is investigated. It is shown that the robust stability of a system over this region is equivalent to the existence of a number of matrix polynomials with certain bounds on their degrees, which satisfy a specific relation. This existence condition can be easily converted to a semidefinite programming (SDP) problem, and handled by means of the available software tools. Moreover, the important case of polytopic uncertainty is studied separately. It is shown that the results provided in [1] and [2] are, in fact, special cases of the ones presented in this paper, when a certain parameter in the given formula is set to zero. Finally, the sampling technique for obtaining an SDP problem from a parametric problem is exploited to further simplify the conditions given for robust stability over the polytope.

Bibliography

[1] R. C. L. F. Oliveira and P. L. D. Peres, "LMI conditions for robust stability analysis based on polynomially parameter-dependent Lyapunov functions," *Sys. Contr. Lett.*, vol. 55, no. 1, pp. 52-61, Jan. 2006.

[2] G. Chesi, A. Garulli, A. Tesi, and A. Vicino, "Polynomially parameter-dependent Lyapunov functions for robust stability of polytopic systems: an LMI approach," *IEEE Trans. Automat. Contr.*, vol. 50, no. 3, pp. 365-370, Mar. 2005.

[3] S. Kau, Y. Liu, L. Hong, C. Lee, C. Fang, and L. Lee, "A new LMI condition for robust stability of discrete-time uncertain systems," *Sys. Contr. Lett.*, vol. 54, no. 12, pp. 1195-1203, Dec. 2005.

[4] M. C. de Oliveira and J. C. Geromel, "A class of robust stability conditions where linear parameter dependence of the Lyapunov function is a necessary condition for arbitrary parameter dependence," *Sys. Contr. Lett.*, vol. 54, no. 11, pp. 1131-1134, Nov. 2005.

[5] V. J. S. Leite and P. L. D. Peres, "An improved LMI condition for robust D-stability of uncertain polytopic systems," *IEEE Trans. Automat. Contr.*, vol. 48, no. 3, pp. 500-504, Mar. 2003.

[6] D. C. W. Ramos and P. L. D. Peres, "A less conservative LMI condition for the robust stability of discrete-time uncertain systems," *Sys. Contr. Lett.*, vol. 43, no. 5, pp. 371-378, Aug. 2001.

[7] A. Trofino and C. E. de Souza, "Biquadratic stability of uncertain linear systems," *IEEE Trans. Automat. Contr.*, vol. 46, no. 8, pp. 1303-1307, Aug. 2001.

[8] P. A. Bliman, "A convex approach to robust stability for linear systems with uncertain scalar parameters," *SIAM J. Contr. & Opt.*, vol. 42, no. 6, pp. 2016-2042, 2004.

[9] P. A. Bliman, "An existence result for polynomial solutions of parameter-dependent LMIs," *Sys. Contr. Lett.*, vol. 51, no. 3-4, pp. 165-169, Mar. 2004.

[10] G. Calafiore and M. C. Campi, "A new bound on the generalization rate of sampled convex programs," in *Proc. of 43rd IEEE Conf. on Decision and Contr.*, vol. 3, pp. 5328-5333, Bahamas, Dec. 2004.

[11] J. Löfberg and P. A. Parrilo, "From coefficients to samples: a new approach to SOS optimization," in *Proc. of 43rd IEEE Conf. on Decision and Contr.*, vol. 3, pp. 3154-3159, Bahamas, Dec. 2004.

[12] J. Nie and M. Schweighofer, "On the complexity of Putinar's Positivstellensatz," *J. Complexity*, vol. 23, pp. 135-150, 2007.

[13] D. Jibetean and E. D. Klerk, "Global optimization of rational functions: a semidefinite programming approach," *Mathematical Programming*, vol. 106, no. 1, pp. 93-109, May 2006.

[14] C. W. Scherer and C. W. J. Hol, "Matrix sum-of-squares relaxations for robust semi-definite programs," *Mathematical Programming*, vol. 107, no. 1-2, pp. 189-211, Jun. 2006.

[15] J. Löfberg, "A toolbox for modeling and optimization in MATLAB," in *Proc. of the CACSD Conference*, Taipei, Taiwan, 2004 (available online at http://control.ee.ethz.ch/~joloef/yalmip.php).

[16] S. Prajna, A. Papachristodoulou, P. Seiler and P. A. Parrilo, "SOSTOOLS sum of squares optimization toolbox for MATLAB," *Users guide*, 2004 (available online at http://www.cds.caltech.edu/sostools).

[17] J. Lavaei and A. G. Aghdam, "A necessary and sufficient condition for robust stability of LTI discrete-time systems using sum-of-squares matrix polynomials," in *Proc. of 45th IEEE Conf. on Decision and Contr.*, San Diego, CA, pp. 2924-2930, 2006.

[18] R. C. Li, "Lower bounds for the condition number of a real confluent Vandermonde matrix," *Math. Comp.*, vol. 75, no. 256, pp. 1987-1995, Oct. 2006.

[19] Z. Chen and J. Dongarra, "Numerically stable real number codes based on random matrices," *Lecture Notes in Comp. Sci.*, vol. 3514, pp. 115-122, 2005.

[20] V. Barthelmann, E. Novak and K. Ritter, "High dimensional polynomial interpolation on sparse grids," *Advan. Comput. Math.*, vol. 12, no. 4, pp. 273-288, Mar. 2000.

[21] J. Lavaei and A. G. Aghdam, "Optimal periodic feedback design for continuous-time LTI systems with constrained control structure," *Int. J. Contr.*, vol. 80, no. 2, pp. 220-230, Feb. 2007.

[22] C. H. Lee, "Solution bounds of the continuous and discrete Lyapunov matrix equations," *J. Opt. Theo. Appl.*, vol. 120, no. 3, pp. 559-578, Mar. 2004.

[23] E. de Klerk, M. Laurent and P. A. Parrilo, "On the equivalence of algebraic approaches to the minimization of forms on the simplex," *Positive Polynomials in Contrl.*, Lecture Notes in Contrl. and Inf. Sci., Springer, vol. 312, pp. 121-132, 2005.

[24] J. B. Lasserre, "Global optimization with polynomials and the problem of moments," *SIAM J. on Opt.*, vol. 11, no. 3, pp. 796-817, 2001.

[25] D. Jibetean and M. Laurent, "Semidefinite approximations for global unconstrained polynomial optimization," *SIAM J. on Opt.*, vol. 16, no. 2, pp. 490-514, 2005.

[26] P. A. Parrilo, "Semidefinite programming relaxations for semialgebraic problems," *Mathematical Programming*, vol. 96, no. 2, pp. 293-320, 2003.

www.ingramcontent.com/pod-product-compliance
Lightning Source LLC
LaVergne TN
LVHW022312060326
832902LV00020B/3415

* 9 7 8 3 6 3 9 0 3 4 1 7 2 *